Economics and Finance for Engineers and Planners

Managing Infrastructure and Natural Resources

Other Titles of Interest

Civil Engineering Practice in the Twenty-First Century: Knowledge and Skills for Design and Management, by Neil S. Grigg, Marvin E. Criswell, Darrell G. Fontane, and Thomas J. Siller (ASCE Press, 2001). Details the essential skills and strategies to supplement the technical preparation of engineers and other technical professionals. (ISBN 978-0-7844-0526-0)

Infrastructure Planning Handbook: Planning, Engineering, and Economics, by Alvin S. Goodman and Makarand Hastak (ASCE Press and McGraw-Hill, 2006). Discusses comprehensively the major concepts, methods, and contingencies involved in infrastructure planning, upgrades, and maintenance. (ISBN 978-0-7844-6003-0)

Infrastructure Reporting and Asset Management, edited by Adjo Amekudzi and Sue McNeil (ASCE Proceedings, 2008). Describes current approaches to asset management and highlights the importance of best practices in infrastructure reporting. (ISBN 978-0-7844-0958-9)

Preparing for Design-Build Projects: A Primer for Owners, Engineers, and Contractors, by Douglas D. Gransberg, James E. Koch, and Keith R. Molenaar (ASCE Press, 2006). A professional reference that covers the basics of developing a design-build project. (ISBN 978-0-7844-0828-5)

Risk and Reliability Analysis: A Handbook for Civil and Environmental Engineers, by Vijay P. Singh, Sharad K. Jain, and Aditya Tyagi (ASCE Press, 2007). Examines the risk and reliability concepts that apply to a wide array of problems in planning, design, construction, and management of engineering systems. (ISBN 978-0-7844-0891-9)

Economics and Finance for Engineers and Planners

Managing Infrastructure and Natural Resources

Neil S. Grigg, Ph.D., P.E.

Library of Congress Cataloging-in-Publication Data

Grigg, Neil S.
Economics and finance for engineers and planners : managing
infrastructure and natural resources / Neil S. Grigg.
    p. cm.
Includes bibliographical references and index.
ISBN 978-0-7844-0974-9
1. Engineering economy. 2. Infrastructure (Economics) I. Title.
TA177.4.G75 2010
   658.15—dc22

                        2009035442

Published by American Society of Civil Engineers
1801 Alexander Bell Drive
Reston, Virginia 20191
www.pubs.asce.org

Cover photograph by Dave Williams.

Contents

Part III: Economic and Financial Tools for Managers

Preface

As engineers gain experience with infrastructure and environmental systems, they realize that deeper management skills are required to complement their technical knowledge. Professional associations and accreditation agencies also recognize this, and thus they have added management topics to their lists of required core competencies. These topics range across many subjects, but economics and finance are central because paying for systems is often the determining factor in decisions.

At Colorado State University, we responded to the need for broader knowledge by revising our undergraduate and graduate civil engineering curricula to introduce integrative topics, and I taught three courses about infrastructure and the environment that covered both management and technical topics. As I developed these courses, I realized that they had recurring themes of public affairs, law, economics, and finance. These themes surfaced in cases of project permitting, infrastructure finance, public involvement, and other scenarios that are familiar to engineers and public works managers.

When preparing these topics, I fell back on my engineering education at West Point, where we had studied broad topics of the social sciences and economics. We were required to subscribe to the *Wall Street Journal*, and we discussed many topics of the era marked by the end of the Eisenhower presidency. Over the years, the *Wall Street Journal* has been a rich resource, and a book has even been written on how to use it to understand economics (Lehmann 1990). I came to realize that economics and finance explain many elements of national and international affairs, including how infrastructure and environmental systems are managed and regulated. A topic from my West Point education that has been particularly meaningful is the economics of national security, which forms a bridge to today's emphasis on critical infrastructure protection.

The disciplines of economics and finance have grown in importance for the civil engineers, construction managers, and public works and utility officials who manage infrastructure and environmental systems. Though they may receive their basic educations in different disciplines, they all require the broad management knowledge specified by the American Society of Civil Engineers in its *Body of Knowledge* (ASCE 2008).

If knowledge of economics and finance was important in the past, it is more so now, in the midst of the turmoil seen in the housing markets, credit markets, stock markets, and government stimulus plans that is unprecedented since the Great Depression. Civil engineers, construction managers, and public works and utility officials have been deeply affected by recent events, and they need a greater understanding of nontechnical issues so they can navigate today's choppy business and government waters.

This book presents the core issues of economics and finance and how they relate to the work of civil engineers, construction managers, and public works and utility officials. It contains more than the basic knowledge on these topics specified within the *Body of Knowledge*. Its primary purpose is to explain how the core issues of economics and finance apply to the management of infrastructure and the environment, and it also explains broad topics to help professionals with higher-level responsibilities and personal advancement in challenging careers in the public and private sectors.

I would like to acknowledge the help I received along the way with this book. First, I think of the many army officers who pioneered in solutions that touch infrastructure problems. I did not have the privilege to meet most of them, but I think of inspirational military leaders such as George Marshall, who oversaw all aspects of the United States' World War II effort; Garrison Davidson, superintendent of West Point in 1957 and Dwight Eisenhower's engineer in the Italy campaign of World War II; and Lucius Clay, who oversaw civil affairs in postwar Berlin. Next, I think of the civilian public works managers I have known over the years. Some of them—people like Jim Martin of Fresno, California, and Myron Caulkins of Kansas City—set a tone of professionalism that has inspired me to this day.

Of course, many educators, economists, and finance specialists have created and published the knowledge that informs this book. Though I do not mention them by name, I have been inspired by leading thinkers who sought to apply the tools of economics and finance to the toughest infrastructure and environmental problems. They developed theories and ran tests that have helped us with everything from the notion of infrastructure services as public goods to the details of utility finance.

I was helped a lot in preparing the book by the ASCE Press staff, notably Betsy Kulamer, ASCE Press editor. She arranged for reviewers, and one of them made helpful comments on the first draft of the book.

References

ASCE (American Society of Civil Engineers). (2008). *Civil engineering body of knowledge for the 21st century*. ASCE, Reston, VA.

Lehmann, Michael B. (1990). *The Business One Irwin guide to using the Wall Street Journal*. 3rd ed. Business One Irwin, Homewood, IL.

Part I

Economics for Infrastructure and the Environment

1

Economics and Finance for Infrastructure and Environmental Management

Management Decisions for Infrastructure and the Environment

As the global population approaches 7 billion, infrastructure and environmental systems are becoming more and more critical to society's future. Their performance is essential to survival, to economic advancement, and to a better quality of life. Although the technologies underlying infrastructure and environmental management are important, the central issues are economic, financial, and political. If this was not clear earlier, it has certainly been clarified by the financial crisis, which was caused by the collapse of a bubble in housing prices.

Managers and engineers can lay the foundation for a sustainable future by creating and sustaining infrastructure systems that meet human needs and protect the environment. You can sense the urgency of this mission across issues ranging from climate change to basic water supply and sanitation. It is critical to daily struggles, such as paying for energy and transportation, protecting public health, and providing safe schools. All these depend on management of infrastructure and the environment.

Effective decisions about infrastructure and the environment involve constructed systems costing trillions of dollars and environmental programs that affect everyone. Managers require tools from the disciplines of economics and finance so that they can effectively ponder consequences, evaluate strategies, and make wise decisions about the balanced use and protection of environmental resources.

Whether infrastructure and environmental leaders start in engineering or another field, their decisions hinge on common skill sets in their required curricula. For example, the American Society of Civil Engineers (2008) has a broad "Body of Knowledge" that requires economics in its social sciences component and finance as part of business knowledge. Other professions with leading roles in infrastructure and the environment also recognize the need for economic and financial tools.

The disciplines of economics and finance draw from the same pool of knowledge, and both use monetary values. However, finance deals with how to pay for things, and economics deals with broader questions about allocating society's resources. For infrastructure and environmental managers, economics is useful for planning and policy analysis, as well as to explain the importance of public decisions. Finance deals more directly with money and is needed to plan, budget, and control the activities of enterprises. Together, the two disciplines make up an indispensable toolkit for infrastructure and environmental managers.

A manager of infrastructure and environmental systems using these tools might be a public works official seeking to finance a capital improvement program. Another user of the tools might be a consulting engineer who must cut expenses on a large transit project with cost overruns. Still another situation where the tools would be useful might involve a development project that requires mitigating flood risk through creative land use plans. These types of public works and environmental programs have an impact on local economies and require engineers and managers to reach into their toolkits and use economics and finance to find answers.

This book explains the issues of infrastructure and the environment in three conceptual areas: the problem space where decisions are required, the occupational spheres of engineers and managers, and the knowledge domains of economics and finance (Fig. 1-1).

Infrastructure and Environmental Systems and Decisions

The intersection of infrastructure and environmental systems occurs when constructed facilities and public services create impacts on the natural environment. The constructed facilities might be housing developments, roads, underground utilities, or other components of infrastructure. The natural environment comprises land, water, and air and also the living things that inhabit it.

The main infrastructure systems discussed in the book are the built environment, energy and water, transportation and communications, and waste management; Chapter 4 provides a classification system for them.

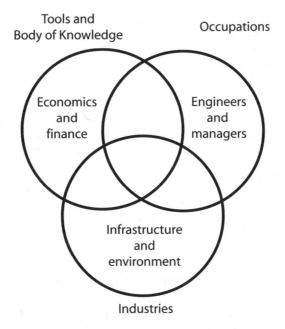

FIGURE 1-1. The convergence of problems, occupations, and knowledge areas

Several systems are discussed in Chapter 7, which is on utility economics, and the construction industry, which cuts across the infrastructure sectors, is discussed in Chapter 8. Natural resources and environmental systems are presented together in Chapter 6.

Infrastructure and environmental managers work in such sectors as public works, utility, and construction management and in such occupations as civil engineering, planning, and public works. The economic sectors in which they work include transportation, construction, energy, natural resources, utility management, water resources, environmental protection, and land development. Though much of their work involves managing money and resources, their technical backgrounds may not include economics and finance. Even if they have training in business management, law, or the social sciences, they still need to learn how to apply economics and finance to infrastructure and environmental decisions.

Engineering students have had high school economics, which is usually part of the social studies curriculum. However, few of them can answer simple questions such as the size of the U.S. budget, gross domestic product, or national debt. And they have not been exposed much to the economic and finance issues of local and state governments.

Managers from different disciplines migrate from one role to another in jobs such as public utility or government manager, consultant, and developer. Their need for economics and finance does not depend so much on their undergraduate discipline as it does on the decisions they face in their jobs. Examples of these jobs include

■ a rising public works manager seeking an executive position in a utility, public works department, transportation district, or city management;

■ a consulting engineer who markets to clients, learns the business side of the industry, and meets with national association staff for public policy work;

■ a professional who migrates from engineering or construction to the business side of land development; and

■ a government executive with a technical background who directs a government or military agency program and/or a regulatory program.

Industries and Occupations of Infrastructure and the Environment

This book uses three occupations as main indicators of where infrastructure and environmental jobs are: civil engineers, environmental engineers, and architects. These three occupations show where the action occurs on the infrastructure and environmental fronts, including the construction industry. The total national employment among these three occupations, as of May 2007, was 405,410.[1] The total employment for the three occupations, aggregated by industry, is shown in Table 1-1. These industry groups account for about 90% (365,760 of the total 405,410 jobs) of the national employment of architects, civil engineers, and environmental engineers. Infrastructure and environmental managers doing similar work would also be clustered in these industries.

The largest employment group is in architecture, engineering, and related services (58% of the jobs). If you add the management jobs to this category, the total reaches 60% of all jobs. The government group includes state, local, and federal governments in that order and accounts for another 22% of the jobs. (Military jobs are not included in these government totals.)

Construction-related jobs are diverse, but they account for only 8% of the jobs. Most civil engineers' work is in nonresidential construction, and architects are employed about 50/50 among residential and nonresidential construction. Therefore, the rapid decline in housing starts would not

---

[1] The employment figures are from data gathered by the U.S. Bureau of Labor Statistics that are accessible from its Web site (www.bls.gov). The data do not vary greatly from year to year.

TABLE 1-1. Employment of civil and environmental engineers and architects, 2007

NAICS industry group	Civil engineers	Environmental engineers	Architects	Total
Architecture, engineering, and related services	126,700	14,980	91,960	233,640
State government	32,300	6,150	1,030	39,480
Local government	28,710	5,070	1,150	34,930
Construction	27,310	190	5,450	32,950
Federal executive branch	9,230	3,850	1,300	14,380
Management, scientific, and technical consulting	4,030	9,650	730	10,380
Total	228,280	39,890	101,620	365,760

Note: NAICS stands for the North American Industrial Classification System, which is explained in Chapter 2.

Source: U.S. Bureau of Labor Statistics (www.bls.gov). Data as of May 2007.

be expected to reduce civil engineering jobs as much as construction jobs directly related to housing.

Most of the remaining 10% of jobs for these industry groups is found among management and administration; remediation and other waste management services; manufacturing; colleges, universities, and professional schools; real estate and land subdivision; mining and energy; and transportation (other than in engineering services and government).

The conclusion from this breakdown is that about 90% of jobs that deal with infrastructure and the environment are in professional services, government, and construction-related work. The topics covered in this book have been selected to focus on the work that occurs in these sectors.

Required Types of Knowledge and Skill Sets

Infrastructure and environmental managers perform similar work, even if their undergraduate educations are in different fields. Figure 1-2 shows the diverse paths they might follow to arrive at their jobs. To gain a perspective on their experiences, I polled a group of them involved with both infrastructure and the environment with the question: "How much economics and finance do public works and infrastructure managers need?"

In general, the respondents to my poll thought that finance was more important than economics. They wrote that economics is a broad subject

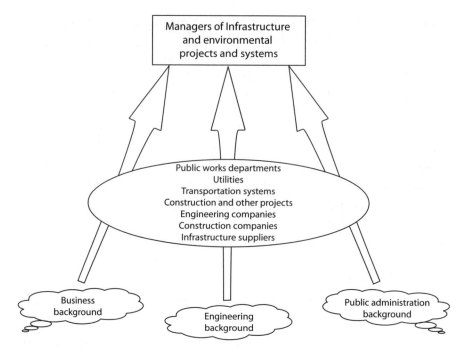

FIGURE 1-2. The diverse paths taken by managers to work on infrastructure and environmental systems

and not appreciated as much for its practical value as is finance. One person responded that economics is useful "to read the newspaper." Although they might not recognize the value of economics, these managers would probably agree that economic topics such as inflation, public sector decision-making, and benefit-cost analysis are useful.

The managers I polled recognized that the undergraduate curriculum is too crowded to offer much economics and finance, and they thought that finance can be learned on the job. The most enthusiastic response from those polled was about the management training course at the Navy Civil Engineer Corps Officer School, which consists of an intensive eight-week practical curriculum covering budgets, finance, contract law, scheduling, inspection, asset/infrastructure management, and leadership skills.

The bottom line is that the managers appreciate practical knowledge that they can apply to their work. Thus, they see the usefulness of finance, but they think economics can be used more to understand issues as part of general education and to make sense of public attitudes and incentives. With finance, they can "follow the money" through the infrastructure cycle.

The Design of This Book

In working with public managers and engineers, you can see many applications of economics and finance to decisions about infrastructure and environmental systems. The design of this book focuses on these decisions and how they require economic trade-offs and the effective use of financial resources. The book does not attempt to present a complete course on either economics or finance, but it seeks to explain the most relevant topics for infrastructure and environmental managers, especially public sector economics and finance and quantitative analysis methods.

After considering the experiences of infrastructure and environmental managers and the industries and occupations in which they work, two sets of topics were selected for the book. For economics, a set of core topics from microeconomics and macroeconomics sets the stage for explaining the economic characteristics of the main industries represented by infrastructure and the environment. These topics include how an economy works to draw on and impact infrastructure and environmental systems, the balance between market and government in infrastructure decisions, measuring the economy and its sectors, supply and demand for natural resources and public services, productivity and its dependence on infrastructure, capital flows and the financing of infrastructure, and economic analysis methods. This set of topics lays the foundation for discussing the main industries in which infrastructure and environmental decisions occur.

Topics within the discipline of economics can focus on systems or on resources—for example, industrial economics, resource economics, agricultural economics, environmental economics, energy economics, and water economics. They can also be classified by scale, such as microeconomics and macroeconomics or urban and regional economics. Part I of the book explains how economic concepts apply to infrastructure and environmental decisions. For example, Chapter 3 explains the incentives for land development that drive the need for infrastructure. Chapter 4 reviews government policies on investments in infrastructure. Chapter 5 explains how the demand for transportation depends on both incentives and regulatory controls. In Chapter 6, the national agenda for environmental regulation is reviewed, and the trade-offs between control and incentives are explained. In Chapter 7, on utility economics, planning for capacity expansion is explained as an economic decision tool. A key issue explained in Chapter 8 is how construction costs depend on inflation, among other variables.

The discipline of finance is about practical subjects such as preparing budgets and reading financial statements. It involves decisions based on the bottom line—profit or loss, the rate of return, and the cost of business. It includes skill areas such as accounting, budgeting, revenue forecasting, cost control, and the analysis of investments. The second part of the book begins with a summary of the field of finance as it applies to infrastructure and the environment. The topics covered in this part are accounting, public finance and budgeting, utility revenue sources, and capital financing and markets. The third part of the book covers tools and analysis methods. It includes asset management, engineering economics and financial analysis, institutional analysis, and the use of economics and finance to promote sustainable infrastructure and the environment.

The content of the book focuses on practical data and methods and not on theory. This is particularly the case for the economics topics, more so than those focused on finance, which is a practical subject anyway. The design of economics part of the book has only one theory chapter (Chapter 2) and six chapters devoted to issues within the infrastructure and environmental sectors (Chapters 3–8). The theory chapter is presented at a general level to introduce topics that are most relevant to infrastructure and environmental systems. In this sense, it does not treat economic theories in depth but explains how they apply to infrastructure and environmental problems.

Reference

American Society of Civil Engineers. (2008). *Civil engineering body of knowledge for the 21st century: Preparing the civil engineer for the future.* American Society of Civil Engineers, Reston, VA.

2

Economics for Infrastructure and Environmental Decisions

Infrastructure, the Environment, and the Economy

The discipline of economics explains how society makes decisions about the allocation of its resources. For infrastructure and environmental systems, these decisions are usually made through combinations of private and public sector economics. Private sector actors make decisions on the basis of their own incentives, and public sector actors take into account the broader public interest.

Several economists who have won the Nobel Prize have blended private and public sector economics to explain infrastructure and environmental decisions. They have probed market and government failures, public sector roles, regulation, and incentives, among other topics. These have even led to subfields of economics, such as public choice theory.

This chapter explains selected basic issues for private and public sector economics and how they influence decisions about infrastructure and the environment. These issues include how infrastructure and environmental systems work within an economy, the balance between market and government in infrastructure decisions, measuring the economy and its sectors, supply and demand for natural resources and public services, productivity and infrastructure, capital financing, and economic analysis methods. The chapter also explains social issues that confront infrastructure and environmental managers and how interest group activity within industrial sectors influences economic outcomes.

The ability of economics to explain a range of things, from personal choices to the global economy, makes it the "queen of the social sciences." Some people think it is dull, and refer to it as the "dismal science," but because of its power to explain how things work, it is not dismal at all.

For example, a basic economics course covers such societal issues as industrial development, poverty, crime, education, and the environment. And there are the different focuses of microeconomics and macroeconomics. On the one hand, microeconomics explains how individual people, firms, and small organizations make economic decisions in the marketplace (Krugman and Wells 2004). These decisions determine consumer demand, price levels, and the supply of goods and services. Microeconomics recognizes that because the market does not always work well, a society may experience "market failure." To address this failure, economists explain how resources can be allocated using nonmarket mechanisms such as government action, regulation, and collective action. On the other hand, macroeconomics, by aggregating economic activity, explains variables such as national income, unemployment, investment and savings, and consumption for regions or whole nations. It addresses larger questions of fiscal and monetary policy, government budget actions, and international economic issues (Baumol and Blinder 2006).

The knowledge base for both microeconomics and macroeconomics is very broad, and the following topics were selected for this chapter to explain the underlying issues of infrastructure and environmental decisions:

■ definition of an economy and how it works;
■ the balance between government and the market in the economy;
■ measuring the economy;
■ supply, demand, and equilibrium;
■ productivity;
■ capital stocks and flows;
■ money and banking;
■ social equity and poverty;
■ international economic issues;
■ economic analysis methods.

Definition of an Economy and How It Works

An economy is the sum of the transactions and economic relationships between the elements of an accounting unit, such as a firm, city, region, state, or nation. It involves the jobs, payments, and trades of the industries and firms, banks and financial institutions, consumers, and government in an economic accounting unit. The myriad small and large transactions conceptually look like Fig. 2-1, which illustrates the flows and circulation of funds through the economy along with the goods and input markets of a basic economic system. In these, businesses sell goods and services to the public, and the public provides inputs in the forms of labor, land, and capital. Businesses pay wages, rents, and interest, which provide consumers with funds to buy the goods and services.

Government can be a quasi-business sector, but it also collects taxes to finance services such as national defense, road building, and aid to education. Government taxation is a redistribution machine. If it takes too much money from either business or the public, there will not be enough left for private market activity.

In addition to government activity, Fig. 2-1 also shows inputs from natural resources and infrastructure assets. Though these provide essential inputs to the economy, they represent different types of capital and are not as easy to include in economic accounts as private business activity.

In the United States, the private market and public activity work together in a "mixed economy." The relationships between them, which are only hinted at in Fig. 2-1, turn out to be very important to infrastructure and environmental management.

The Balance Between Government and the Market in the Economy

A large share of the economy is devoted to construction, reconstruction, the use of natural resources, utilities, and other issues of infrastructure

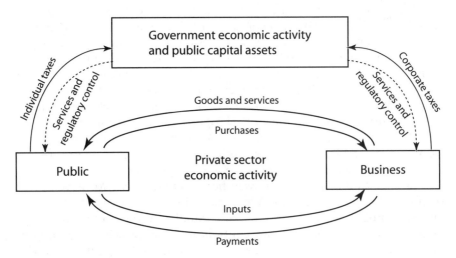

FIGURE 2-1. Flows in the economic system

and the environment. How decisions are made about them requires choices between private and public sector approaches. For example, the privatization of infrastructure systems entails a choice for capitalism and against socialism. Environmental regulation also illustrates this kind of choice. Should it be done using a government "command-and-control" approach, or should the market be used to allocate environmental resources?

Capitalism, Socialism, and Central Planning

Although private enterprise dwarfs government activity in most sectors, government actions in infrastructure and environmental management are very significant. These may involve the ownership of infrastructure and/or resources, as well as regulation and control.

Economists search for ways to use the market for infrastructure and environmental management, but many decisions remain closer to socialism than to capitalism. Under socialism, decisions about the allocation of resources are made by communities of people and by the government rather than by the competitive marketplace. When the government coordinates economic decisions, it practices a form of central planning.

Most countries practice a mixture of capitalism and socialism, and the relative magnitudes of the two systems measure which one is predominant. Political debates about them go on continually, and decisions about infrastructure and the environment are at their center because they involve the public's business.

The United States is considered a capitalist nation, but it has flirted with socialism in the past. Much of the legislation that affects health and welfare, such as construction safety, began during periods of social emphasis. For example, after the Soviet Revolution in 1917, some people thought that central planning had advantages over Western capitalistic systems. Under central planning, the government exercises command and control over the economy to decide what people need and what to produce.

During the Great Depression, these theories were tested in the United States. Franklin Delano Roosevelt's New Deal agencies, such as the Public Works Administration and the Works Progress Administration, oversaw vast infrastructure programs, including the building of major structures such as Hoover Dam. The Tennessee Valley Authority was also created in this period, and many people still consider it an ongoing experiment in socialism.

The pendulum has swung away from experiments with government economic activity such as these. Whereas, during the 1930s, the government turned to public works investments to stimulate the economy, in recent years it has relied on interest rate controls and even on the distribution of money to stimulate consumer spending. This seemed to change in 2009, with the giant federal stimulus package approved in an attempt to reverse severe economic decline.

Welfare economics is a niche of economics that focuses on public decisions to maximize social welfare. To use it analytically, you must know how to measure the public's social welfare, and this theory is at the heart of concepts like the "triple bottom line" of sustainability or the balancing of the social, economic, and environmental outcomes from public decisions.

Utility theory has similar goals. According to it, a person's utility is a measure of satisfaction from some outcome, and economic behavior can be explained by attempts to increase utility. It can be used, for example, to determine the values of coefficients in social welfare functions. These theories can be useful, but people have different ideas about public choices and their welfare. Decisions about problems such as clean water or transportation involve many facets of public welfare. People must decide how much to invest in them. These decisions are as much in the realm of politics as in economics.

Valuing public benefits is not precise because we lack exact knowledge of public preferences and because decisions are referred to the political process, in which decisions receive the scrutiny of voters and political leaders. A branch of economics called "public choice theory" addresses how the public makes decisions. Another field that addresses these same topics is "political economy."

Since socialistic approaches and government planning have been shown to be flawed, most countries have sought some form of competition. Still, when the government determines "needs," as it does for highway or wastewater plant construction, and then allocates funds to help meet those needs, it is engaging in a form of central planning. So the bottom line is that for infrastructure and environmental systems, the nation is still searching for ways to introduce more competition and less command and control into its decisions.

Government Actions in the Economy

The federal, state, and local levels of government have huge impacts on the economy, especially in the infrastructure and environmental sectors. State and local governments have an even greater influence over infrastructure and environmental decisions than the federal government. Table 2-1 shows the magnitudes of economic activity of the three levels of government, which include 1 federal government, 50 state governments, and 87,525 local governments, including special districts (U.S. Bureau of the Census 2004).

Government policies have great leverage over decisions about infrastructure and the environment. Many of these begin with federal legislation, which cascades in its effects on state and local governments. Given its large

TABLE 2-1. Economic activity of the three levels of government

Measure of activity	Federal	State	Local	National total
Units of government	1	50	87,525	87,576
Gross domestic product (GDP)				$13.2
Government share of GDP	$0.52	$1.12[a]		$1.64
Total employment (nonfarm)				134,000,000
Government employment	2,713,000	5,111,000	14,290,000	22,114,000

Note: Dollar amounts are in trillions.

[a]Combined state and local government.

Sources: The data for the end of 2006 are from the interactive tables at the Web sites of the Bureau of Labor Statistics (http://www.bls.gov) and the Bureau of Economic Affairs (http://www.bea.gov).

size, the federal government is able to influence national economic performance through fiscal and monetary policy. In addition, the government's fiscal and regulatory tools give it policy clout over sectors of the economy.

Chapter 4 will present a six-system framework for public infrastructure. This list illustrates examples of government policy areas for each system:

■ *the built environment:* public buildings, housing finance, tax exempt bonds, regulation of health and safety;
■ *energy:* energy policy through legislation, public power generation, power regulation through the Federal Energy Regulatory Commission and public utility commissions;
■ *water:* federal water policy, dam safety, government water projects, the regulation of health and safety;
■ *transportation:* federal highway legislation, support for transit, rail subsidies, airline regulation, Intracoastal Waterway maintenance, transportation security;
■ *communications:* regulation of frequency spectrum, the Federal Communications Commission, the deregulation of telecommunications; and
■ *waste management:* the regulation of solid, hazardous, and nuclear wastes, and environmental controls.

As a macroeconomic policy tool, the federal budget strongly influences both the national and global economies. Its revenues, expenditures, and deficit or surplus determine economic outcomes, including the level of the national debt. Unfortunately, the federal budget is heavily committed to social spending, national defense, and debt service. This leaves less funding for programs such as infrastructure and the environment, as shown by Fig. 2-2.

In addition to passing laws and regulating the economy, the government provides many services itself. Though some could be provided by the private sector, others are clearly appropriate for government agencies. Table 2-2 shows examples of government and private activity in six infrastructure sectors.

Public housing is an example of a government service area. It is a difficult program area because good faith attempts to provide basic housing have sometimes created dysfunctional housing areas that are difficult to manage. Therefore, the public housing program continues to be reexamined.

Government's function as regulator is very important in infrastructure and environmental management. Each infrastructure sector is regulated or guided by federal statutes for health, environment, safety, and other purposes. Examples of statutes for different infrastructure sectors include:

- transportation: Transportation Equity Act;
- water: Safe Drinking Water Act;
- energy: Federal Power Act;
- the built environment: Occupational Health and Safety Act;
- waste: Resource Conservation and Recovery Act; and
- communication: federal communications acts.

Continuing calls for "regulatory reform" mean "Get the government off our backs," but you also hear calls for more regulation. The level of regulation expresses the balance point in the public's desires.

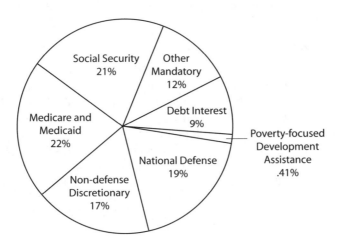

FIGURE 2-2. Distribution of the U.S. budget

Source: Learner 2007.

TABLE 2-2. Examples of government and private activity in six infrastructure sectors

Sector	Government service	Private service
Transportation	Public roads	Airline industry
Water	City water supply utility	Private water company
Energy	Public power (e.g., Tennessee Valley Authority)	Private electric utility
Built environment	Public buildings	Commercial buildings
Waste	City solid waste collection	Private waste hauler
Communications	Internet security	Telephone service

Government borrowing to finance deficit spending is an important instrument of national economic and fiscal policy. Government debt was some $10.8 trillion in early 2009, or more than $30,000 for every citizen (U.S. National Debt Clock 2009). With the huge stimulus package of 2009, this seems to be headed upward. Interest expenditures in the federal budget are directly related to this debt, most of which is in Treasury bonds and bills. How government borrowing takes place is covered in Chapter 13, which discusses the capital markets.

The infrastructure deficit is actually a "third deficit" of the nation. In addition to the national debt of over $10 trillion, the nation also has an inter-generational obligation to pay future Social Security benefits. Infrastructure funding needs also represent a massive obligation for future generations.

Finding the Balance

In an ideal capitalist world, all activity would be by the private sector and the public sector would disappear. Though the private sector does meet most of society's needs through the "invisible hand" described by the eighteenth-century economist Adam Smith, it does not always meet social needs. When this happens, it constitutes "market failure."

In an ideal socialist world, the private sector would disappear. However, government and community activity can also fail. Government activity is not subject to market competition and can be like the failed central planning system of a communist country. Nor is public sector activity always in the public interest. Sometimes public sector actors make decisions based on self-interest.

In the real world, we have a mixture of private and public sector activity. For example, your airline ticket is issued by the private sector, but your passport and airport security are handled by the government.

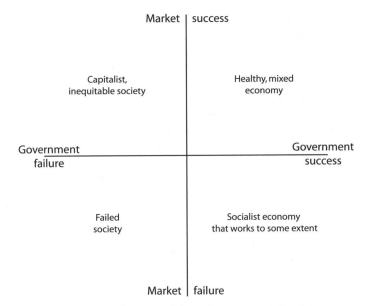

FIGURE 2-3. Market and government success and failure

As shown in Fig. 2-3, market and government economic activity can be illustrated by a matrix of successes and failures. A society where the market and government both work is a successful, mixed economy (like the United States). If the economy works but government fails, you have a capitalist but inequitable society (like a small country run by a strongman). If government works but the market fails, you have a socialist country. Small Scandinavian states are considered models of this type of socialism, but they are more mixed than purely socialist. Finally, if both government and the market fail, you have a failed society.

The balance between government and private provision of infrastructure services changes with time. Engineering designs may be produced in house by government agencies or, more commonly, outsourced to engineering firms. Most construction is by private contractors, but some public agencies have construction and/or remediation arms. Even activities such as software development can be in the public sector if done by government or public university providers.

In the mixed U.S. system, three methods of delivering public services are in use: the traditional public model, privatization, and a mixed model of managed competition. In the traditional model, public services are provided by government agencies. Privatization means turning over services completely or partially to the private sector. Managed competition is a compromise that allows government services or privatization or a mixture (Greenough et al. 1999).

Sentiment in the United States and in other countries is tilting toward using the private sector to deliver services whenever possible. This is an example of the need for a mixed economy, but one that prefers the private sector whenever it makes sense.

All three approaches to delivering public works services offer benefits and challenges. Management trends such as "reinventing government" and "budgeting for outcomes" influence practices (Osborne and Gaebler 1992; Osborne and Hutchinson 2004). Also, in addition to business and government, public interest work and philanthropy fill many of society's needs as they operate in the area between the market and government.

Measuring the Economy

Economic Indicators

The discipline of economics requires many quantitative indicators to measure the economy. These enable us to keep track of business cycles, employment, productivity, and other variables. Two issues illustrate the need for valid indicators. One is whether investment in infrastructure provides a good stimulus to the economy, and another is how you account for environmental resources within the economic product of a nation. Both of these are addressed by ongoing debates about how to improve economic metrics.

A starting point for measuring the economy is to classify it by dividing it into industries, which are clusters of economic activities that have similar production and exchanges. Classifying industries is complex because there are many combinations and levels. The North American Industrial Classification System (NAICS) provides a framework, but many combinations of industries cut across its basic categories. For example, the "transportation industry" includes everything from roads through ocean shipping.

Industries that account for much of the work of infrastructure and environmental managers fall into the categories for utilities (NAICS 22); construction (NAICS 23); professional, scientific, and technical services (NAICS 54); and public administration (NAICS 92). Public administration covers government activity, which is treated like an industry under the NAICS (U.S. Bureau of the Census 2008b).

The basic yardstick of economic activity is gross domestic product (GDP), which measures the sum of the production in the nation as recorded by the U.S. Bureau of Economic Affairs (2008). The United States has the world's largest GDP, which in 2008 reached nearly $14 trillion in goods and services per year, or just over $45,000 for each of the nation's 300 million citizens. (These GDP estimates are in current dollars and reflect continuing inflation.)

Productivity is another important economic metric because it is the key to improving the standard of living. It measures how much you are producing relative to your investments or inputs. Debates occur as to how much infrastructure investments help national productivity and whether environmental regulation is too much of a drag on productivity. Measuring productivity is complex, and there are no simple answers to these questions.

Productivity statistics are compiled by the U.S. Bureau of Labor Statistics (2008), which publishes indexes of labor and multifactor productivity for economic sectors and manufacturing industries. However, the bureau does not keep productivity statistics on the construction industry or utilities, making it difficult to assess their impact on the economy (Building Futures Council 2005).

Analysis of the effects on productivity of infrastructure investments and environmental regulation is complex and requires research studies, and even then the answers are not always clear. In Chapter 4, this is illustrated with a summary of a national conference on infrastructure economics that sought to explain links between infrastructure investment and productivity.

Economic modeling can show the influences of input variables on economic outcomes. For example, an input-output model showing interactions among industries that was developed by a Nobel laureate economist, Wassily Leontief, has been used to show the influence of infrastructure investments on economic productivity. But this model requires a great deal of data and is not used much. Modeling technologies have advanced, however, and other econometric models can now be used to study the same phenomena. Today, models such as IMPLAN are available for economic impact modeling. This model creates social accounting matrices and models of local economies (Minnesota IMPLAN Group 2008).

Supply, Demand, and Equilibrium

In economic theory, the concept of "supply and demand" is useful to locate the equilibrium point of a balance between the sellers and buyers of goods and services. At this equilibrium point, the balance between supply and demand determines the price and the quantity sold. For infrastructure work, however, demand is not driven by price so much as by needs or expectations. The demand for infrastructure generally means the load placed on a highway, a water treatment plant, or another system. In turn, the demand for infrastructure commodities, such as lumber and steel, does determine their prices.

In market economics, supply refers to the quantity of a good or service that a producer is willing to provide at a certain price. In infrastructure work, it means furnishing the system required to meet the demand placed on it by the public. For example, the supply of a highway system is determined by the capacity of the road system that is built. Again, the supply of the input quantities, such as asphalt or guard rails, is determined by the market for those components.

The concepts of supply and demand do not apply directly to environmental resources because the resources are not supplied by industries and the goal is not so much to balance demand with supply as to sustain the resources and limit their use to a renewable rate.

Although traditional concepts of supply-demand equilibrium are not used directly in infrastructure systems, they can explain some important issues. To see this, consider the curves in Fig. 2-4, which represent downward-sloping demand relationships and upward-sloping supply relationships. These curves show that as the price drops, consumers will demand more of

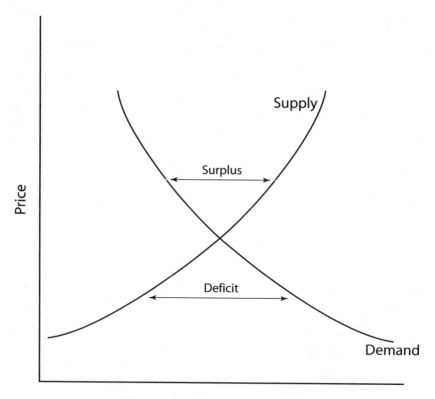

FIGURE 2-4. Supply and demand relationships

a good or service and that as the price rises, producers are willing to supply more of it. For example, consider gasoline as a good. If its price is lower, people want bigger cars, take more trips, and so on. If the price goes up, they want more fuel-efficient cars and more will take the bus instead of drive.

If the quantity demanded increases substantially with a drop in price, demand is said to be elastic. If it does not, it is inelastic. A downward-sloping demand curve can be understood intuitively by considering that when a good or service becomes cheaper, people will demand more of it. Take lumber, for example. If it costs less, more will be demanded for construction. By the same token, as the quantity demanded goes up, the producer is willing to provide more for higher prices. Thus, supply-demand curves and their equilibrium point reflect a balance between downward-sloping demand and upward-sloping supply. These supply-demand concepts hold generally for theoretical situations where there is perfect competition, but in actual situations, the relationship is more complex. For example, the demand for drinking water is not very elastic, meaning that as its price changes, demand might not change that much.

Although supply, demand, and equilibrium concepts apply more to marketplace situations than to infrastructure and environmental decisions, they help us to see and explain the need to supply infrastructure services. Take water supply as an example. A minimum quantity is necessary for survival, and people will pay a lot for it. Data show that poor people will expend a large share of household income to buy bottled water when no other sources are available. Therefore, for lower quantities, the price can rise on a relatively inelastic demand curve for drinking water. This means that you do not choose the supply of water strictly on the basis of consumer demand; you must provide basic quantities as a human right for life. This issue is discussed in more detail in Chapter 7 under the topic of utility economics.

Capital Stocks and Flows

Both infrastructure and environmental systems involve what economists call "stocks and flows." Stocks are quantities of something in storage, and flows are changes in them during time increments. Stocks of money, for example, include money in savings, and a flow of money would be the amount of a budget to be spent in one year. Infrastructure is "capital stock," and environmental assets can also be considered stocks. Changes, such as the deterioration or loss of environmental assets, are flows.

The infrastructure investment debates are about how much capital stock the nation can afford and how to pay for it. Table 2-3 outlines the quantities and shows the distribution between public and private stock and types of stock in the United States. Almost all the government stock is in structures,

TABLE 2-3. Inventory of the U.S. public and private capital stock of infrastructure and equipment

Item (2005 data)	Total (trillions of $)
All fixed assets	37.3
All government stock	7.9
All private stock	29.3
Private nonresidential equipment	4.7
Private nonresidential structures	8.8
Private residential housing	15.8

Source: U.S. Bureau of the Census 2008a.

and state and local governments own most of it. The value of private housing has declined during the financial crisis of 2008–9.

For an overall view, consider that if the total value of U.S. built facilities was $30 trillion, new construction was $1 trillion annually (see Chapter 8), and facility lifetimes were 30 years, the nation would just stay even with the need to renew its infrastructure. This does not account for expansion, but it gives a rough balance between the stock and the construction of infrastructure. The actual picture is much more complex. Some facilities last more than 30 years, and the value of expansion is offset to some extent by depreciation.

Another major category of stock is financial assets. These are difficult to estimate, but data on them are recorded in the Federal Reserve's Flow of Funds accounts (U.S. Bureau of the Census 2007). Most stock or wealth is held by households and nonprofit organizations. These are followed by nonfinancial businesses, banking and financing corporations, government and government enterprises, investment companies, and pension funds and insurance companies. The total as of 2005 was about $115 trillion. Obviously, with the tremendous decline of the stock market in 2008–9, this value has declined substantially.

The liabilities against these financial assets are not known exactly. Nevertheless, financial wealth is greater than fixed asset wealth. We also do not know the exact liabilities against infrastructure, in the form of government debt and the deterioration of facility conditions in the form of deferred maintenance.

In addition to infrastructure and financial assets, environmental and other assets include standing timber, fish and game, collectors' items such as art and cultural objects, national monuments, subsoil assets (natural resources), patents, copyrights and goodwill, and human capital. These would have to be offset by liabilities of various kinds (Goldsmith 1982).

A country's net worth can be computed if you take into account all assets and liabilities. This could be massive—such as unfunded Social Security entitlements or the loss of oil reserves—or it could be smaller, such as holding assets in the form of foreign currencies that are subject to exchange rate fluctuations (Traa and Carare 2007).

Compiling a national balance sheet to show wealth is a complex thing, but Goldsmith (1982) showed the following approximate distribution of assets in the United States: land 10.7%, reproducible tangible assets 29.3%, monetary metals 0.5%, and financial assets 59.5%. The tangible assets category includes all capital stock, but in addition it includes timber, inventories, livestock, consumer durables and semidurables, collectors' items, and research and development equity. Subsoil assets such as oil and gas are included in the land category, which shows 2.0% agricultural land, 6.9% other land, and 1.8% subsoil assets. Using the same ratios as in 1975, land would be about $10 trillion, reproducible tangible assets about $30 trillion, and net financial assets about $60 trillion. The tangible assets category includes all capital stock.

The concept of wealth can be used to classify those who are "wealthy" and those who are living in "poverty." This concept can also apply to use of GDP per capita as an indicator of average wealth and poverty among nations. The United States and other Western nations, along with Japan, have high ratios of GDP per capita. Poor countries, like many in Africa and Asia, have very low ratios.

Money and Banking Systems

Credit for Infrastructure Construction

Credit is essential to stoke infrastructure construction through up-front capital and long-term financing. For both of these, the supply of funds to build infrastructure depends on the money supply, interest rates, and the banking system. During the financial crisis of 2008–9, there has been a problem with credit for private and public borrowers.

Money is a medium for exchange or trade, a standard of value, and a store of value through savings. It can be backed by a commodity like gold or silver (representative money), or it can be money because the government says it is money (fiat money). Because fiat money is not backed by a commodity, it is only valid as long as people trust the guarantee of government (Federal Reserve 2006).

The money supply is at the heart of the price-setting mechanism of a nation's economy. It is a critical determinant of inflation, which affects business strategy. If money is unlimited, then its value approaches zero and you have inflation. If money is too scarce, it is hard to get and businesses cannot expand or operate by borrowing.

Today, in addition to bills and coins, the money supply includes checking accounts and bank deposits. Credit cards seem like money, but they create debt and are not, strictly speaking, part of the money supply.

Early in its development, the United States followed the gold standard and only issued as much money as it had gold to back it up. This limited the supply of money and restricted access to credit for small business owners and farmers. Today, the money supply is no longer supported by gold but depends on stable fiscal and monetary policy.

The Banking System

The U.S. banking system manages the flow of money and furnishes credit to finance investments and operations of all kinds of enterprises, including infrastructure organizations. Banking has changed dramatically, and distinctions between banks and other financial institutions are becoming blurred. The banking system includes the Federal Reserve System and various types of banks—national banks, state banks, commercial banks, savings banks, savings and loan associations, credit unions, and investment banks. Commercial, national, and state banks are the largest sources of loans to small businesses. The Office of the U.S. Comptroller of the Currency (2005) supervises the national banking system, which includes some 2,000 national banks.

A central bank is a government-sponsored or -chartered bank that serves to stabilize the money supply and to undertake other essential functions of monetary policy for a country. The central bank of the United States is the Federal Reserve System, with its 12 regional banks. The Fed is controlled by a 12-member Board of Governors, with a chairman appointed by the president. By setting the interest rate at which it lends money to its member banks, the Fed determines the supply of credit. The equilibrium interest rate depends on the supply and demand for money.

Today, the Fed has two roles: to ensure financial stability and to promote a balanced economy. For financial stability, it uses its discount window and the discount rate, which is the rate at which banks can borrow funds from the Federal Reserve to maintain their required reserves. To promote economic balance between inflation, expansion, and recession, it uses its federal funds rate, which is the interest rate at which banks lend their balances at the Federal Reserve (called federal funds) to other banks overnight.

Throughout the 1990s, the Fed used monetary policy to control inflation and manage the economy. Monetary policy uses the tools of open market operations, discount policy, and reserve requirements. As this was written in 2009, the Fed had lowered interest rates and pumped money into the banking system to ease liquidity due to the housing market crisis and stock market decline. In the future, the Fed will face challenges such as deregulation, technological changes in payments systems, globalization, and mergers and acquisitions in financial services (Federal Reserve 2006).

Private investment banks are important for infrastructure financing because they assist companies and government agencies to raise money by selling securities in the capital markets. They also act as intermediaries in trading financial instruments. The large U.S. investment banks include Salomon Smith Barney, Morgan Stanley, and Goldman Sachs. The regional and smaller investment banks may have niches such as bond trading. The investment banks came under terrific pressure during 2008 and 2009, and some large ones had to close operations.

As an example of how investment banks operate, Goldman Sachs is divided into three segments: investment banking, trading and principal investments, and asset management and securities services. Its Investment Banking Division handles mergers, acquisitions, divestitures, and issuance of equity or debt capital. Its Trading and Principal Investments Group focuses on various investment instruments such as fixed income, currencies and commodities, and equities. Its Asset Management Group offers mutual funds and other products and services for high-net-worth individuals (Goldman Sachs 2006).

A development bank is a special kind of lender that provides funds for economic development and infrastructure development. A state infrastructure bank (such as the California Infrastructure and Economic Development Bank) can be a government mechanism for assembling various revolving loan programs into a single activity. The World Bank makes loans for development on a worldwide basis. Regional development banks, such as the Inter-American Development Bank and the Asian Development Bank, have similar missions.

The International Monetary System

With an interconnected global economy, ratios of currency values take on great importance and determine interest rates for construction and other investments. Currency exchange rates are set through the "international monetary system," and they involve not only market forces but also political actions. The international monetary system stems back to World War II, when it was evident that a new system was needed, and in 1944, at Bretton Woods, New Hampshire, an international conference led to the establishment of the International Monetary Fund and the International Bank for Reconstruction and Development (World Bank). Today, the World Bank is very active in investments and actions related to both infrastructure and the environment.

Industrial Economics and the "Iron Triangle"

Industrial economics is a branch of economics that studies specific sectors or industries. In these, the relationships among business, government, and interest groups are important, and the "iron triangle" is a useful concept to explain them (Fig. 2-5). The term is used by political scientists to explain relationships between legislatures, the bureaucracy, and interest groups, and it can also help to explain incentives and economic behavior among these actors.

In an iron triangle, interest groups such as businesses, unions, and trade groups may extend electoral support to Congress in return for friendly legislation and oversight. In turn, the interest groups may advocate for the bureaucracy by lobbying Congress to get less regulation and special favors. The bureaucracy gets funding and political favors from Congress in return for policy choices and support. During the 1950s, President Dwight D. Eisenhower worried about the growth of the "military-industrial complex," but today the

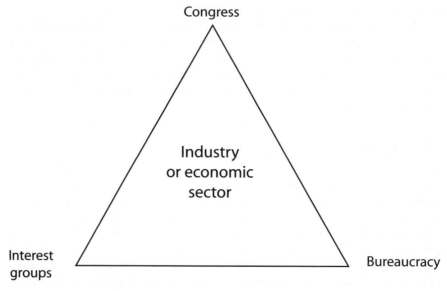

FIGURE 2-5. A general view of an industry's iron triangle

term applies as well when interest groups lobby for their favorite projects. In subsequent chapters, the iron triangle concept will be used to show how various infrastructure and environmental industries function.

Social Equity Issues and Poverty

Providing infrastructure services for lower-income people is an important mission around the world. The level of income is a measure of relative wealth among individuals, families, regions, and nations. It determines the relative amounts of infrastructure a village or nation can afford. Clearly, there is a close relationship between income and level of infrastructure development.

Economic metrics focus on efficiency and equity. Economists like to explain that these two concepts mean how big the pie is and how it is sliced. Thus, efficiency is a measure of total production, and equity deals with the distribution of the income gained from this production and the resulting social benefits. For instance, when household income is low, families may live in poverty and be deprived of amenities or even the basic necessities of life.

Economics provides ways to measure poverty, such as the "Lorenz Curve," which is a graph of percent income as a function of percent population. If the curve is a straight line, you have equality, but if it dips below the line, you have relative degrees of inequality. Coefficients can be calculated to demonstrate the relative inequality existing in a society.

Poverty and social equity are important issues for infrastructure and environmental systems because they offer both the necessities of life and important amenities that improve the quality of life. For example, a transit system can provide mobility so that low-income people can get to jobs. An urban recreation area can provide opportunities for inner-city kids to participate in field trips and improve their educations. In developing countries, access to potable water and sanitation is a life-or-death matter that depends on income levels. Taking account of social equity is an important requirement for infrastructure and environmental decisionmakers.

International Economic Issues

Globalization is becoming more important because international trade and economics affect infrastructure and environmental quality in many ways. One example is the global construction market, which includes the work of consulting engineers. Another example is the work of multinational corporations, which may exploit natural resources in the countries where they do business. Still another example of an international issue with strong links to infrastructure and environmental quality is institutional poverty.

Infrastructure and the quality of the environment are life-or-death matters in developing countries, which have exploding urban populations, debt crises, and serious social problems. After World War II, the United States–sponsored Marshall Plan was a critical foreign aid program to get Europe back on its feet. Today, providing water and sanitation services in developing countries is a key public health issue for billions of people. Traffic congestion and adequate infrastructure are holding back social progress in many cities of developing countries. Though international development and aid are important factors to improve the environment and quality of life in poor countries, emphasis is moving toward market-based solutions and away from sole reliance on direct assistance and aid from governments and multilateral institutions. Nongovernmental organizations also have big roles. Financial mechanisms such as microloans are able to bridge many gaps because they unleash the creative capacity of people, rather than making them dependent on handouts.

Methods of Economic Analysis

A wide range of economic analysis methods are in use, from basic equations and graphs to complex simulation models. Of particular interest to infrastructure and environmental managers are those that enable us to compare alternative ways to invest public funds. These methods derive from basic concepts such as compound interest and the time value of money, and evolve into more sophisticated multiobjective decision models.

Chapter 16 explains how, although some of these techniques are called economic analysis methods, they are the same as financial analysis methods. The exception might be the economic analysis methods that take into account social, environmental, and economic accounts that do not involve finance, strictly speaking.

One widely used method is benefit-cost analysis, which enables us to identify and quantify the full range of positive and negative contributions of a potential course of action. These can then be compared to determine which among several possible courses of action is best. The challenge of applying this kind of decisionmaking method is to correctly quantify the benefits and costs.

Although, strictly speaking, benefit-cost analysis and other economic analysis methods belong more to economics than to finance, they are explained in Chapter 16. As Chapter 16 delineates, the reason for not presenting them here is to avoid confusion among similar concepts and methods. Chapter 6 also explains a key issue in water economics: valuing water in its various uses. This challenge must be confronted so that we can find efficient ways to allocate water among users.

References

Baumol, W. J., and Blinder, A. S. (2006). *Macroeconomics: Principles and policy.* 10th ed. Thomson/SouthWestern, Mason, OH.

Building Futures Council. (2005). Building Futures Council urges Bureau of Labor Statistics to establish construction productivity measures. http://www.thebfc.com. Accessed May 8, 2006.

Federal Reserve. (2006). Fed 101: History of the Federal Reserve. http://www.federalreserveeducation.org/fed101/history/. Accessed July 4, 2006.

Goldman Sachs. (2006). Client services. http://www.gs.com/client_services/. Accessed July 4, 2006.

Goldsmith, R. W. (1982). *The national balance sheet of the United States, 1953–1980.* National Bureau of Economic Research, Cambridge, MA.

Greenough, G., Eggum, T., Ford, U. G., III, Grigg, N. S., and Sizer, E. (1999). "Public works delivery systems in North America: Private and public approaches, including managed competition." *Public Works Management and Policy, 4*(1), 41–49.

Krugman, P., and Wells, R. (2004). *Microeconomics.* Worth, Upper Saddle River, NJ.

Learner, M. (2007). Budgeting-for-justice. Bread for the World. http://www.bread.org. Accessed August 2, 2007.

Minnesota IMPLAN Group. (2008). About us. http://implan.com. Accessed June 21, 2008.

Office of the U.S. Comptroller of the Currency. (2005). *A guide to the national banking system.* U.S. Government Printing Office, Washington, DC.

Osborne, D., and Gaebler, T. (1992). *Reinventing government: How the entrepreneurial spirit is transforming the public sector.* Addison-Wesley, Reading, MA.

Osborne, D., and Hutchinson, P. (2004). *The price of government: Getting the results we need in an age of permanent fiscal crisis.* Basic Books, New York.

Traa, B., and Carare, A. (2007). "A government's net worth." *Finance & Development, 44*(2), 46–49.

U.S. Bureau of Economic Affairs. (2008). http://www.economicindicators.gov/. Accessed May 7, 2008.

U.S. Bureau of Labor Statistics. (2008). Productivity and costs. http://www.bls.gov/lpc/home.htm. Accessed May 8, 2008.

U.S. Bureau of the Census. (2004). *2002 Census of governments, vol. 1, Organization.* U.S. Government Printing Office, Washington, DC.

———. (2007). *Statistical abstract: Table 1148—Flow of funds accounts, 1952 to 2005; Table 700—Top wealth holders with net worth of $1 million or more—*

Number and net worth by state, 2001. U.S. Government Printing Office, Washington, DC.

———. (2008a). *Statistical Abstract: Table 701—Net stock of fixed reproducible tangible wealth in current and real (2000) dollars, 1925 to 2005*. U.S. Government Printing Office, Washington, DC.

———. (2008b). 2002 North American Industrial Classification System Codes and Titles. http://www.census.gov/epcd/naics02/naicod02.htm. Accessed May 7, 2008.

U.S. National Debt Clock. (2009). http://www.brillig.com/debt_clock/. Accessed February 25, 2009.

3

Urban and Regional Land Use

Land Development and Demand for Infrastructure

The driving forces for the demand for infrastructure and environmental impact are land development and use, including both urbanization and the use of rural land for commercial and recreational activities. Although urban areas occupy less than 3% of U.S. land, their wealth and business activities dominate the economic landscape. These built environments comprise the organizing framework for infrastructure and public services in urban areas, and they have large impacts on the environment. Agriculture, forestry, resource development, and outdoor recreation also have large impacts on the environment and create demand for infrastructure.

Microeconomics explains how local land uses are planned and developed by entrepreneurs, and their decisions have significant impacts on other players. Regulatory controls affect everything from the size and cost of streets in subdivisions to the provision of social goods such as affordable housing.

This chapter explains the economic issues that determine how urban and regional land development and use occur. Chapter 6, which covers the economics of natural resources, addresses the economics of land use in a broader sense and describes land development activities in rural areas.

Economics considers social and behavioral forces as well as flows of money, and it draws on other disciplines to address urban and regional development. The interdisciplinary arena of urban and regional economics focuses on jobs, housing, transportation, and public services. A field called regional science seeks to place these in a unified framework to explain how transportation and land development influence regional patterns. Regional science involves economics, planning, sociology, civil engineering, political science, and architecture, among other disciplines (Isard 2003).

A Model of the Built Environment

Chapter 1 explained how infrastructure systems comprise inputs and flows to and from the built environment, which is itself a physical system with links and nodes. Land uses and structures are represented by these nodes, and the arteries for transportation, communication, and flows of commodities are represented by these links. Figure 3-1 illustrates these nodes and links that constitute the physical basis for urban economic activity.

The nodes of the built environment are visible on a map. They are the clusters of structures within built-up areas that have economic purposes. Portions of them are separated by open spaces, and economic activities occur within the subnodal areas, such as zones, subdivisions, and other areas within the larger nodes. Transportation models, for example, represent traffic flows from one zone to another.

A classification of land uses and nodes of urban structure can be derived from the U.S. Bureau of the Census's tabulation of construction spending

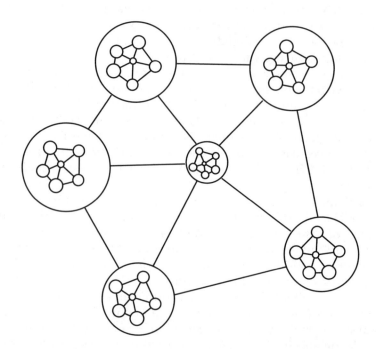

FIGURE 3-1. Nodes and connecting links in the built environment

(see Chapter 8). Links are infrastructure arteries that carry flows of transportation, communications, and utility commodities. Roads, streets, and rail links are the connectors and foundation for land uses. Their importance goes beyond the function of conveying traffic to include their roles as shared space for utility corridors, social interaction, and disaster response. Above-ground and underground utility networks distribute water, energy, communications, and information. Solid wastes are transported by truck. The economic category of "transportation and utilities" is a convenient way to organize the links.

The built environment is in a continual cycle of growth and decline. As the United States evolved from its agricultural roots to become a highly developed nation, it fought a Civil War and passed through the Industrial Revolution before 1900. Then it entered the automobile era and a period of expanding federal government role. The eras of postwar suburbanization and road building helped shift the stage to environmentalism and regulated development (Tarr 1984). Today, problems of infrastructure and environment are shaped by an information-dominated society with high energy costs. Meanwhile, complex forces continue to shape and reshape the built environment.

Urban and Regional Economics

Land Uses, the Economic Base, and Jobs

Within an urban area, the economic base determines the supply of jobs. Basic or export jobs produce goods and services for export out of the region, and service jobs produce goods and services that are consumed locally. Examples of export jobs are those in basic manufacturing, in consulting services for other regions, and at the headquarters of a trade association. Examples of service jobs would be those of retail workers, local insurance agents, small local construction companies, and local restaurants.

Looking at a local economic base this way explains why officials are so eager to attract new, basic jobs. With new jobs come new businesses and homes, property may go up in value, and the economy becomes more vibrant. Additionally, construction jobs increase during growth phases. The job base must stay strong or the community will stagnate. If a city lacks export jobs, it cannot simply thrive on local service jobs because there is no generation of money to buy products made outside the area. The economic doctrine of comparative advantage applies to urban regions just as it does to nations. A region has an advantage and exports to other regions, which provide it with the income it uses to purchase outside goods and services. The United States is a large economic area, and thus its regions can export to each other, but it has a national current account deficit with some other nations, and internationally it imports more than it exports.

If the distribution of jobs in an area remains about the same, then the ratio of total population to the number of export jobs also remains about the same. For example, one export job may support a population of six people, more or less. If you expect an increase in export jobs, you can project the corresponding increase in population. By the same token, you will get an increase in service jobs to go along with this increase in export jobs. Using fixed ratios like this is called the export base technique for economic forecasting.

Land uses require public services and transportation for firms making decisions about where to locate and to create jobs. The stocks of land and buildings determine the valuation of an area and its tax base. The availability of services affects the cost of doing business and the attractiveness of business sites. Location decisions determine employment and growth, which make possible the improvement of life in the community.

The Urban Housing Stock

Residential housing comprises the majority of construction spending and is a strong driver of infrastructure demand. Housing is classified by density and includes low to high densities, planned unit developments, estate areas, and mobile home areas. A city's housing stock will normally include a range of buildings that vary in age, condition, value, and style. The quality of the housing stock is also an important indicator of economic health and the ability to finance growth and improvement through the tax base.

With the exception of public housing, homes are provided by the private sector's housing industry. The public sector has important roles in regulating housing and supporting it with infrastructure development. Providing affordable housing is an important issue in communities, and local governments sometimes require developers to provide a mix that includes affordable housing.

The regulation of land development and housing is normally done by the planning and zoning departments of local governments, which set performance standards by issuing codes and subdivision regulations that follow state and federal requirements as well as local rules. Flood insurance is one example of a federal program that influences local land use controls. Regulation has important influences on the availability, quality, and cost of housing.

Most homeowners borrow capital to build or buy a home. Debt plus regular operations and maintenance expenses are repaid from current income, and a mortgage payment will normally have four components: principal, interest, taxes, and insurance. If no credit for mortgages was available, home ownership would be difficult and limited to the wealthy. Therefore, to assist homeowners, the federal government and some states have developed housing finance agencies. In 1934, only 40% of households owned homes. By 2001, the homeownership rate was 68% (U.S. Department of Housing and Urban Development 2007).

The United States has agencies called Fannie Mae and Freddie Mac to help make mortgages available. They have portfolios of more than $1.4 trillion and are regulated by the Office of Federal Housing Enterprise Oversight. These quasi-government agencies are supposed to step in to buy mortgages when others are not buying them (Hargerty 2006). During the housing crisis that began in 2007–8, these agencies got into serious financial trouble and had to be bailed out by the federal government. Thousands of banks, hedge funds, and institutional investors also buy mortgages from the lenders that originate them.

Holding the cost of housing down is an important but difficult goal due to the scale of providing infrastructure and facilities and complying with codes and regulations. During the 1960s and 1970s, the U.S. Department of Housing and Urban Development sponsored programs for innovation in housing and to build new cities. Generally, these were not successful due to institutional problems.

The large sums required to finance home ownership can lead to distortions in the economic system. During the 1980s, the nation had a savings and loan crisis, during which many savings and loan banks failed. In 2008, the "subprime" mortgage crisis occurred because too many lenders had made risky loans.

Given the large economic stakes, the politics of housing are intense. The iron triangle for the industry has much of the construction industry at one corner, government agencies at three levels at another corner, and politicians with stakes in housing at the third corner. Interest groups for housing range from business interests, such as the National Association of Home Builders, to advocates for low-income people, such as Habitat for Humanity.

As an antipoverty program, public housing has been difficult for urban areas. Public housing projects can be successful if the right principles are followed, for example, when tenants take responsibility for their own problems. Globally, about 1 billion people live in poverty in unhealthy settlements, which often lack organized law and order. New slums, called "informal settlements," are without government sanction, basic services, or a legal basis for home ownership (Peirce 2007). Here, shelter is a matter of life and death, and creating and financing it are a big problem. Barriers include the shortage of capital, lack of trunk infrastructure, land administration, rent control, and poor planning. The World Bank gives priority to methods that improve access to housing, but around the world there are still many large pockets of poverty with people living in desperate and vulnerable conditions.

Types of Land Uses

Development statistics on land uses are maintained by the U.S. Bureau of the Census (2006) under its data for construction spending. Principal categories include residential areas; office, commercial, and government buildings; retail areas and amusement and recreation centers; hospitals, schools, and churches; industrial areas and warehouse districts; public safety facilities; and transportation, communication, and utility nodes.

Residential Areas

Residential construction is the biggest piece of the economic pie, and it generates demand for infrastructure services, such as utility services and transportation trips. A vibrant urban area will have affordable housing for varied income levels. Types of residential housing include a range of single-family and multifamily buildings and mobile home parks.

Residential areas are the places where more than 300 million Americans live. Given the large size of the housing industry, much of the construction industry's employment is devoted to it. Also, a large industry provides home improvements, furnishings, maintenance, and other needs of residential areas.

Office, Commercial, and Government Buildings

Office, commercial, and government buildings provide the venues for much of the official life of a city's built-up area. They represent concentrations of important activities and carry some security risks, as was shown by the 1995 Oklahoma City terrorist bombing. They form the core of many cities and are also in developed zones in fringe or edge areas. Modern developments mix these official buildings with residential units, but in some areas, the office, commercial, and government centers become deserted at night.

Public buildings usually are under the responsibility of departments of "buildings and grounds" or the "physical plant." These terms are giving way to the more functional term "facilities management."

Facilities management, or the total management of building space, is receiving more attention as a functional area of management. More and more, smart and integrated buildings link building management, information management, and telecommunications. They can focus on energy saving, control of climate and lighting, security, fire protection, centralized data management, teleconferencing, and telecommunications (Honeywell 2005).

Given their variety, office and commercial centers are subjected to more regulatory controls than those for housing. Large management organizations, such as the U.S. General Services Administration, manage huge portfolios of buildings. Financial sources for commercial real estate are different than for housing finance. They focus on corporate financial sources and investment lenders. Insurance companies and pension funds may own office buildings and centers, even leasing them to government tenants.

Office, commercial, and government centers require active management to remain competitive and vibrant as revenue-producing assets. When complexes with millions of square feet of space deteriorate, large-scale renovations and changes are needed, with corresponding effects on infrastructure. If the areas are allowed to slip into decline, they can become blighted areas. Thus, the renewal of office and commercial centers is an ongoing issue in many urban areas.

Retail Areas and Amusement and Recreation Centers

Retail areas, such as downtowns and shopping centers, generate great demand for infrastructure and services, and they have large concentrations of people and economic activity. Amusement centers, including large facilities such as Disney World, are diverse and create their own needs. Like manufacturing plants, these centers create jobs, but they tend to be lower-wage service jobs rather than higher-wage basic jobs. Depending on the level of activity, the impact of large centers can be substantial, considering pavements, runoff, traffic generation, and waste disposal.

Hospitals, Schools, and Churches

Hospitals, schools, and churches are important activity centers that may require special protection and utility services. In particular, health care facilities are extensive and critical components of the built environment, and they will become even more prominent as the population ages. These facilities may require special attention in infrastructure planning due to the loads they place on basic services, the fact that many do not pay taxes, and their needs for security and protective services.

Industrial Areas and Warehouse Districts

Factories and warehouses are often located in special zones. They can have their own substantial infrastructures, comprising roads, utilities, and energy systems, which need to be coordinated with public systems. Manufacturing plants sometimes have big environmental impacts. On the positive side, they can be job creation mechanisms, and infrastructure authorities may give them concessions to attract the jobs.

Public Safety Facilities

Public safety facilities, including firehouses, police stations, and emergency response centers, are important parts of the urban area and require special coordination with transportation and communication systems. Though the facilities themselves do not have special needs for water or electricity, it is important that they be secure because they are needed to improve security for the rest of the area and facilities.

Transportation, Communication, and Utility Centers

Like public safety centers, transportation, communication, and utility centers should be protected and given special treatment for their roles in providing services to the rest of the area. These can include facilities for water supply and treatment, wastewater treatment, electric power generation, natural gas relay stations, transportation hubs, communication nodes, and solid waste transfer stations. These nodes provide critical infrastructure services, and they represent the connecting points of services provided through utility networks. These networks require large percentages of the capital requirements of infrastructure. For example, water distribution systems are buried assets that comprise around 60% to 70% of the total assets of a water utility.

Urban and Regional Planning and Development

Planning Process and Land Use Control

The development of the nodes and links of the built environment is initiated through urban and regional planning. Raw land is developed or urban areas are redeveloped through infilling and renewal, including brownfield development. A brownfield, as defined by environmental law, is a parcel of land that has had some kind of environmental problem that requires cleanup before the land can be used again. Greenfields are new areas for development without the problems of brownfields.

Land use is mostly controlled by local governments in the United States, not by state or federal governments—except in special cases. Although planning has many checks and balances, it does not always occur through a linear process with clear procedures, because trade-offs are made and the political process has a strong influence on outcomes.

The key to a city's planning process is the set of plans that make up its "comprehensive plan" or the master plan that includes all other plans. These can be divided into sector and area plans. In an ideal process, proposals for development are evaluated for consistency with the comprehensive plan. The developer complies with the plan and pays development fees, and growth occurs according to the plan. However, things are seldom that tidy due to political, economic, and legal forces. A city's ability to enforce land use controls has limits.

The Land Development Process

The process that creates new developments can be represented by a "development game," which comprises the activities undertaken by developers, regulators, and other players from concept to the completion of a land development (Fig. 3-2). It is like a competitive game because it has the procedures, rules, players, referees, and the other elements of a game. Through this game, checks and balances are in play to work collectively toward reaching the shared goals of communities.

Much of the action in the development game can be explained by the incentives of its players. Those shown in Fig. 3-2 include:

- The developers and real estate interests and investors who profit from land development. They are the driving force for development.
- Regulators and politicians, comprising the city government staff, the members of the planning and zoning commission, and the city council. They control the throttle through regulatory authority and decide if development proposals are in the public interest.
- The community and general public and interest groups.
- Planners, architects, engineers, and lawyers—professionals who profit from the fees derived from development work.
- The contractors and suppliers who profit from development work.

The incentives in the development game are shown in Fig. 3-3. The developer's profit motive is the strongest force that drives development. In most cases, the developer will obtain land, get zoning, prepare a plan that will work, get approvals, build infrastructure, and then sell, build, and develop further. The

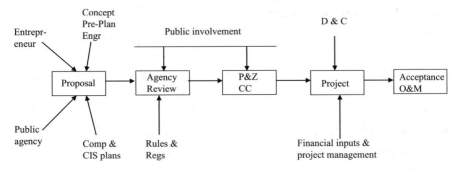

FIGURE 3-2. Steps in the development game

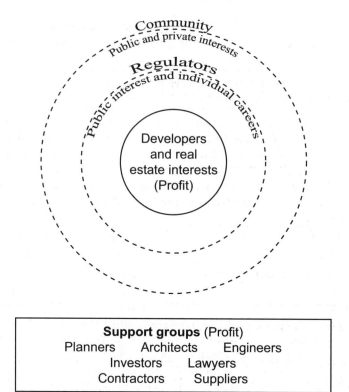

FIGURE 3-3. Incentives in the development game

community has a mixed set of incentives about development. Some win, some lose. Real estate and business interests normally win. Citizens who like the status quo may see themselves as losing from it. Support groups, such as design engineers, contractors, and bankers, have strong incentives to promote development. Regulators are motivated by the public interest, political ambitions, and career advancement.

Development should be controlled by instruments such as city policies, the comprehensive plan, sector plans, and area plans. These plans form a hierarchy:

1. *comprehensive plan:* a master integrated, or comprehensive plan for the jurisdiction;
2. *facility master plans:* master plans for each service or facility category;
3. *needs assessment:* needs assessments linked to the budget process; and
4. *capital programs:* programs to improve each category of infrastructure.

As Fig. 3-4 shows, these plans start at the top with the comprehensive plan of a jurisdiction. The comprehensive plan is a multisector plan that integrates

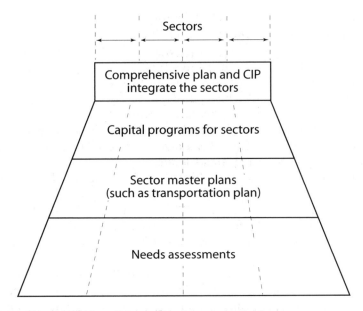

Sectors

Comprehensive plan and CIP integrate the sectors

Capital programs for sectors

Sector master plans (such as transportation plan)

Needs assessments

FIGURE 3-4. Hierarchy of city and infrastructure plans

considerations of land use, services, transportation, utilities, and the rest. Its credibility rests on good needs assessments for the sectors, followed by master plans for the sectors and the capital programs for each sector. The capital programs come together in the jurisdiction's overall capital improvement program.

Citizens have diverse interests in the development process. Generally, they want a desirable community, a good housing mix, a functional transportation system, access to services and employment, and amenities to create a better living environment. They may oppose those development proposals that they perceive as threatening some aspect of their life—for example, with too much traffic, locating undesirable activities near them, and environmental degradation. This gives rise to the well-known "not-in-my-backyard" or "NIMBY" phenomenon.

Development generates attractive business opportunities for the planners, architects, engineers, lawyers offering professional services and for the contractors and suppliers who provide facilities and equipment. Thus, development has its own iron triangle of mutual interests made up of developers, business interests, and public sector participants who benefit from it in one way or another.

To illustrate the development process and the developer's financial incentive, consider this hypothetical example, which I use to explain to students how the development process works. John Jones, an entrepreneur, wants to develop a mixed-use parcel of 160 acres. His company, New Century Properties (NCP), acquires raw land for $20,000 per acre, or $3.2 million (including transaction fees). NCP opens a line of credit with High Risk Capital, LLC (HRC) at 10%. HRC also takes a 25% equity stake, in consideration of sharing some of the risk. Repayment will begin after two years.

NCP hires a landscape architect firm for $150,000 to prepare conceptual plans and submissions to the city. After six months, the conceptual plans are submitted, and this is followed by negotiations and more negotiations. Approval is granted in month 23. NCP must allocate 40% of the land for public uses, and the rest can be sold as single-family lots, at five to the acre.

With approval in hand, NCP hires BC Engineering (BCE) to design the improvements. The estimated cost for infrastructure improvements is $10,000 per acre, including streets and utilities, making a total cost of $1.6 million. BCE gets 15%, or $240,000, for its design and construction supervision work. Engineering is completed in month 30, and this is followed by construction and inspection.

NCP has to pay the city an impact fee of $5,000 per acre for the 160 acres for plant investment and another lump sum fee of $500,000 for street oversizing. The total NCP investment so far is about $6.5 million, not including interest payments to HRC. Construction is completed in month 36.

Pacific Rim Realtors agrees to sell the lots for a commission of 5%. Custom Home Builders also buy lots at discounts of 40% from retail. Sales begin, and all lots are sold by month 84.

Adding in the accumulated interest charges of about $500,000, and considering other NCP expenses, the total investment is now about $8 million. If 60% of 160 acres can be sold as lots at 5 lots to the acre, then 480 lots can be sold for an average of $100,000 each. However, half the lots were sold to home builders at a 40% discount, so NCP's total revenue will be about $38 million.

All the lots are sold by month 84. Realtors get 5% of the revenue, or about $2 million, so the net revenue to NCP is $36 million. Given the total investment of about $8 million, the net profit is about $28 million. HRC gets its 25% equity stake, or $7 million, and NCP ends up with $21 million. The three principal NCP partners worked full time on this for seven years with few paydays. Their payouts are $7 million each, making the average pay about $1 million per year for each partner.

Again, this is just a hypothetical example, but it serves to illustrate some of the principal cash flows, risks, and distribution of costs for land development. The economic impact of land development is large. When you add in the real estate and home improvement industries, you see that the total constitutes a significant share of gross domestic product.

To further pursue the subject of land development, planning information is available from groups such as the American Planning Association. The

National Association of Home Builders offers publications on decent and affordable housing, estimating home construction costs, housing economics, residential codes, residential streets, smart growth, and green building. For developers, the Urban Land Institute offers publications on real estate development, making smart growth work, shopping centers, planned communities, and different types of housing. Other interest groups also generate publications on topics such as smart growth.

Tools for Planning

Tools for urban and regional planning include plans and instruments to direct and control land use decisions, population forecasts, economic forecasts, and land use models.

Instruments to Direct and Control Land Use

Urban areas have an arsenal of tools to direct and control their land uses. These include requirements, codes, and standards such as densities, street standards, open space requirements, drainage standards, utility codes, fire protection requirements, environmental controls, and impact fees.

In addition to outright controls, government can be a participant in land development. Tools might include land acquisition, development rights, and easements; exactions, dedications, land banking, and land trusts; annexation; growth management and growth rate tools; open space and agricultural land protection; and incentive programs.

Population Forecasts

Population studies are used to analyze demands for infrastructure and public services, but population cannot be predicted accurately very far into the future. Population growth is caused by natural change and migration. Natural increase reflects birthrates and death rates. Migration results from people's decisions to move from one area to another. Countries like Japan and Russia are experiencing low birthrates and low migration, and Russia has a high death rate. The United States has a relatively low birthrate, but its population continues to grow because of migration.

Some regions of the world have high rates of urban growth caused by high birthrates and migration. For many years, people have sought to escape rural poverty by moving to cities. Providing infrastructure for this growth is a formidable challenge, especially in megacities, many of which have serious problems of illegal settlements.

States and regions of the United States have large variations due to migration. The Sunbelt has had a long run of growth, while the Northeast has been mostly static. Western energy states have experienced booms and then busts.

The effects of social change on population can be significant in a typical infrastructure planning period of 20 to 50 years. In the United States, a 50-year planning period from 1915 to 1965 would include two world wars, the Great Depression, the New Deal, the Korean War, the Vietnam War, and many technological changes—all still in the memories of people living today.

The basic equation for population change is relatively simple. This formula expresses the compounding of population growth:

$$P_t = P_0(1 + G)^n$$

where P_t is population in a future year t, P_0 is population in the base year, G is the annual growth rate (including natural increase and migration), and n is the number of years.

The effects of the population's compounding can be dramatic. Consider a community with a base population in 2000 of 100,000. Projecting its population with four alternative growth rates yields the numbers given in Table 3-1. Also note, as shown in Fig. 3-5, the huge difference that a growth rate of 3% to 5% can make even after 20 years.

Migration is impossible to forecast with any certainty over the long term. On this point, a newspaper editorial titled "Why Demographers Are Wrong Almost as Often as Economists" explained how demographers missed the postwar baby boom, baby bust, surge of women in workplace, sudden drop in death rates, exodus to the Sunbelt, and other major population changes (Otten 1985).

TABLE 3-1. Example of the effects of compounding on population, using four alternative annual growth rates

Year	No growth	1% growth	3% growth	5% growth
2000	100,000	100,000	100,000	100,000
2005	100,000	105,100	115,930	127,630
2010	100,000	110,460	134,390	162,890
2020	100,000	122,020	180,610	265,330
2030	100,000	134,780	242,730	432,200
2040	100,000	148,890	326,200	704,000
2050	100,000	164,460	438,390	1,146,740

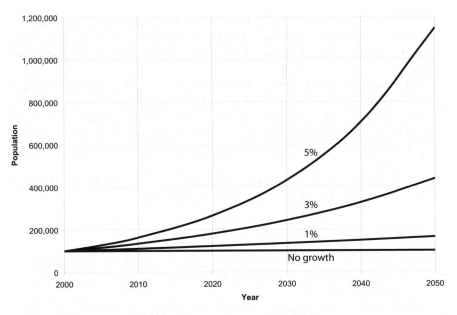

FIGURE 3-5.   Population growth with different rates

Economic Forecasts

Economic studies can analyze the conditions that attract jobs and population growth and create demand for infrastructure. Jobs are classified as basic, or export jobs, and nonbasic, or service jobs. An example of an export job is a manufacturing position. Most infrastructure and utility jobs are in the service categories.

Export jobs drive economic development, and there is normally a relatively stable ratio between export jobs in an area and other parameters of population, including service jobs and total population. If an area lacks export jobs, there will be no flow of money to buy products made outside the area. To see this, consider a farm family where no one has an outside job. The family's members make deals for services from each other, but no one has any money to buy products or services from outsiders. Even if the family is self-sufficient, its standard of living will be low.

The tool for doing economic studies of an urban area using these facts is called the "economic base technique." To apply this technique on a simple basis, you would study a local economy. If the ratio of the total population to export jobs was 5:1, then you could predict that on a very approximate basis, the addition of 1,000 new export jobs would increase the population by 5,000. This is a simple and approximate method. Other economic models consider more variables and thus can perform analyses in greater detail.

Land Use Models

By combining population and economic forecasts, you can to some extent forecast changes in land use. However, because changes in population and economic growth are uncertain, the rate of change of land uses is also uncertain. To improve forecasts, you can add other factors of growth and institutional change to the mix, such as land use controls.

An ideal land use model would examine use patterns and economic forces, and would predict future patterns of growth. If successful, it could create financial advantages and the ability to predict the need for infrastructure. Models like this exist, but they are not able to project land uses very far into the future.

Simulation models for cities and regions can also be developed by combining population, economics, and land use. These are interesting for gaming and might be useful in infrastructure planning, especially for transportation systems.

One type of model is based on the "systems dynamics" technique developed at the Massachusetts Institute of Technology by Jay Forrester (1969). Since Forrester's original work, many variations and software platforms have been developed. Forrester's urban dynamics model simulated three interdependent subsystems: business, housing, and labor. His techniques have been widely applied. During the 1970s, we applied them to the business operations of a water supply utility by simulating the utility's financial and water stocks (Grigg and Bryson 1975). This model had four subsystems: the population and business subsystem, which drove demand for water and revenue forecasts; the water stocks subsystem, which showed the supply and availability of water; the facility subsystem, which showed the quantity and condition of capital facilities; and the water rates subsystem, which allowed for planning for revenues and adjusting charges. Today, a model like this can be created on a spreadsheet for financial and facilities planning.

Geographic information systems, along with databases, can be used to display characteristics of land uses and help in forecasting future patterns.

Disasters and the Built Environment

Given its concentration of people and facilities, the built environment is vulnerable to natural and human-caused disasters, which have enormous economic consequences. The nation's largest recent disaster was the result of Hurricane Katrina, which destroyed a large part of New Orleans. The attacks of September 11, 2001, illustrated the nation's vulnerability to terrorism. Other major recent natural disasters have included the Great Mississippi River flood of 1993, the 1994 Northridge and 1995 Kobe earthquakes, and

Hurricane Floyd in 1999. Providing security against these kinds of threats is the job of government at all three levels, and the critical infrastructure protection programs of the Department of Homeland Security are explained in the next chapter.

References

Forrester, J. (1969). *Urban dynamics.* MIT Press, Cambridge, MA.

Grigg, Neil S., and Bryson, Maurice C. (1975). "Interactive simulation for water system dynamics." *Journal of the Urban Planning and Development Division, ASCE,* May.

Hargerty, J. (2006). "Regulator signals compromise over role of Fannie, Freddie." *Wall Street Journal,* July 5.

Honeywell. (2005). Integrated building solutions: A total capability. http://content.honeywell.com/uk/integbuild.htm. Accessed January 11, 2005.

Isard, W. (2003). *History of regional science and the Regional Science Association International: The beginnings and early history.* Springer, New York.

Otten, A. L. (1985). "Why demographers are wrong almost as often as economists." *Wall Street Journal,* January 29.

Peirce, N. (2007). "World's slum dwellers: More like us than we think." Washington Post Writers Group, http://www.napawash.org/resources/peirce/peirce_8_12_07.html.

Tarr, J. A. (1984). "The evolution of the urban infrastructure in the nineteenth and twentieth centuries." In *Perspectives on urban infrastructure,* ed. R. Hanson. National Academy Press, Washington, DC.

U.S. Bureau of the Census. (2006). Construction spending: Construction at a glance—2005 total. http://www.census.gov/const/www/c30index.html. Accessed May 8, 2006.

U.S. Department of Housing and Urban Development. (2007). Federal Housing Administration. http://www.hud.gov/offices/hsg/fhahistory.cfm. Accessed August 16, 2007.

4

Civil Infrastructure Systems

Infrastructure and Its Subsystems

Although people intuitively understand the abstract concept of infrastructure, they have a better feeling for tangible systems such as roads and water lines. Thus, the concept of infrastructure is a framework for a high-level system of the lifelines and structures that support our social and economic life. Subsystems such as roads and bridges, water pipes, and electric power generating plants are components of this high-level system.

Though the infrastructure concept can seem abstract, it provides a useful framework for discussing broad economic issues such as investment and development. Although interest groups identify more with its subsectors than with infrastructure itself, it is useful for policy studies at the aggregated level. For example, in the U.S. House of Representatives, the policy committee is called the Transportation and Infrastructure Committee. In the Senate, it is called the Environment and Public Works Committee, which has a Transportation and Infrastructure Subcommittee. We need to remember that the actual work of planning, design, construction, and operation takes place within subsectors such as roads and water systems, and you have to be sure to explain what you mean by infrastructure.

Infrastructure is a multifaceted system that aggregates systems with different attributes. It supports economic development, which in turn generates the capital to build and manage infrastructure. Economic development and the quality of life depend directly on structures and lifelines for energy, water, communications, transportation, and waste disposal.

Figure 4-1 shows how infrastructure supports the built environment. The infrastructure systems in turn place loads on the natural environment (e.g.,

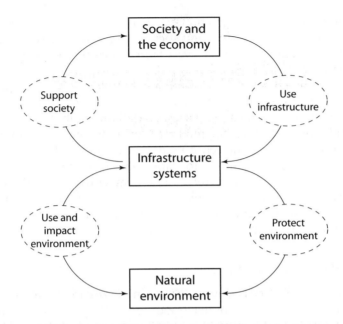

FIGURE 4-1. Infrastructure, the economy, and the environment

runoff from highways), but they also mitigate the environmental effects of built facilities with mechanisms such as wastewater treatment and the use of sustainable designs.

Recognizing that infrastructure is a high-level system, this chapter addresses the economic concepts that are related to it. One concept explains infrastructure as an organizing framework and provides a vocabulary for its analysis. A second concept provides a classification system and a conceptual model for infrastructure. Next, the chapter presents a summary of economic conclusions about the infrastructure "crisis." Three performance concepts for infrastructure management are also explained: life cycle management, infrastructure condition curves, and performance metrics. These are needed to explain how infrastructure affects the economy and how we measure the demand for and supply of it.

Infrastructure Definitions and Classification

A Definition and Classification System for Infrastructure

Infrastructure definitions can be general or specific. On a specific basis, the systems discussed in the book—the built environment, transportation, communications, water, energy, and waste management—involve nodes and links that support the economy and society with necessary services. When you aggregate these to the general level of infrastructure, the selection of subsystems becomes more arbitrary.

As an example of a general definition of infrastructure, the Department of Homeland Security (DHS) included multiple economic sectors in its definition of "critical infrastructure" (DHS 2006). Some of these—such as agriculture and food, public health, and banking and finance—are not physical infrastructure systems in the sense used in this book. DHS subsequently added the phrase "key resources" to clarify what it means. Figure 4-2 compares DHS's critical infrastructure/key resources systems with the physical infrastructure addressed in this book. This expansive definition by DHS illustrates that, because infrastructure is a high-level systems concept, it can mean different things to different groups.

A Model of Infrastructure Systems

Like a human body, infrastructure is a "system of systems." It has links (like roads) and nodes (like cities or neighborhoods). This network terminology describes how built environments in cities and towns are connected by links between vital nodal activities.

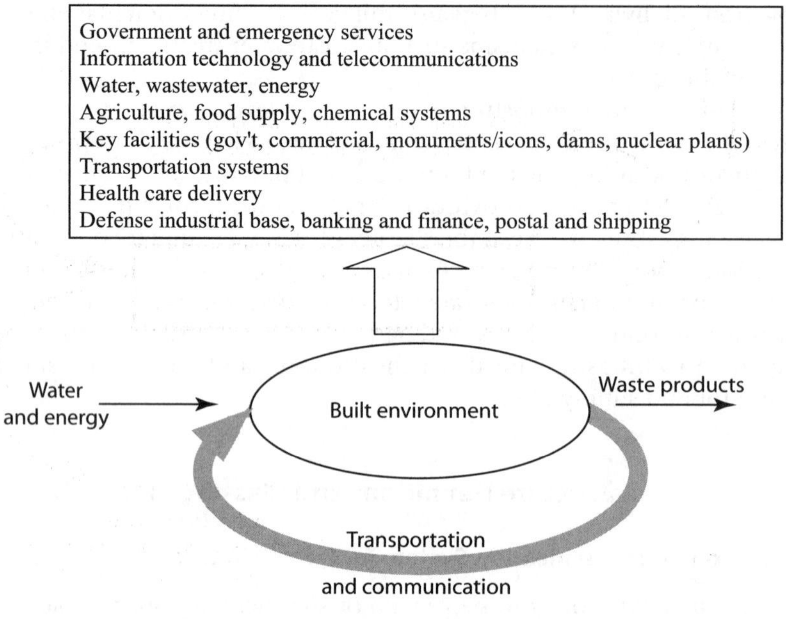

FIGURE 4-2. Infrastructure systems supporting critical systems

Figure 3-1 showed the intercity and intracity links between and within built environments. Cities are separated by rural areas or city-suburb complexes and are connected by networks for transportation, communications, water, and energy. In the model of Fig. 3-1, the built environment is a node and the five other systems are links (water and energy supply systems, transportation, communications, and a waste management system). This concept is easy to remember because it relates the built environment to its flows (transportation and communication), its basic inputs (water and energy), and waste management facilities (Fig. 4-3). Overlap occurs in the model because the built environment is served by the link systems, but the link systems are also part of the built environment. For example, streets are part of the built environment and the transportation linkage system.

As shown by these examples, a classification system for infrastructure will show sector, ownership, and level:

- sectors and subsectors (e.g., water, transportation, housing);
- ownership (e.g., public, private, military); and
- levels (e.g., local roads as a level of surface transportation).

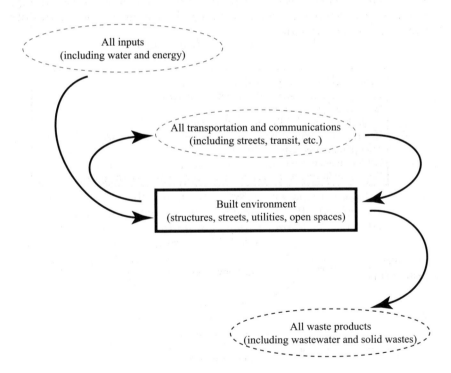

FIGURE 4-3. Infrastructure systems to support the built environment

The civil infrastructure system is a conceptual framework developed by the National Science Foundation (1993) when its infrastructure program was initiated. It explained that critical infrastructure systems are "the constructed physical facilities which support the day-to-day activities of our whole society, and provide the means for distribution of resources and services, for transportation of people and goods, and for communication of information."

Infrastructure sectors at the second level constitute large and complex industries unto themselves. Sectors such as transportation and energy have their own government agencies, regulatory laws, and trade associations. The fourth level is still complex enough to be considered as a systems level and to have its own textbooks and annual specialty conferences. The lowest level shows components that are designed and built, such as a water main. As an example of levels of infrastructure systems, you can readily identify five layers: a water main is part of a distribution system, which is part of the water supply sector, which is part of the water resources field, which is addressed by the civil infrastructure field. Another example is shown by Fig. 4-4, which uses a transportation system to illustrate the hierarchy of infrastructure systems.

Public Choices for Infrastructure

A Review of the Infrastructure "Crisis"

In recent decades, infrastructure has been in the headlines because of its deteriorated condition or lack of funding. The American Society of Civil Engineers' Infrastructure Report Card is a way to keep this situation in the public's eye. Beginning about 1980, infrastructure condition became a hot-button issue (Choate and Walter 1981). There were vigorous debates about investing in transportation, water, energy, communications, the built environment, and waste management systems. These debates have not ended, but they have moved off the front page as the nation faces many other financial challenges, including military commitments, Social Security, and a current account deficit. Infrastructure issues return to the front page when there is a failure such as a bridge collapse, but then they die down as larger issues overtake them.

FIGURE 4-4. Example of the hierarchy of infrastructure systems

We have learned a lot from these debates. National cover stories about infrastructure highlighted the problems and showed the needs to invest trillions of dollars in infrastructure renewal. The earliest debates were about how to finance infrastructure in general, but our attention has since turned to sector-specific needs and financing methods.

The debates were about how to allocate society's resources to infrastructure systems versus other public goods—such as health care, defense, and welfare. These questions are answered mostly by government decisions about tax rates and public budgets. The issues are complex. For example, what parts of infrastructure constitute public goods that require investment versus private goods that should be supplied by the market? In a transportation system, for example, choices about public issues, such as when to resurface a road, are clearly different than private choices, such as the type of automobile to buy.

Is There a Shortfall in Public Capital Investment?

A number of years ago, a conference of economists and other scholars experienced with infrastructure issues examined the question "Is there a shortfall in public capital investment?" (Munnell 1990). They noted that during the 1980s more output had gone to consumption than to public investment, and they confirmed that public capital stock had been neglected. They discussed government fiscal problems and the extent of underinvestment in public infrastructure, along with its economic consequences and policy prescriptions.

These economists at the conference agreed that public capital investment plays an important role in the quality of life and economic activity and that declines in public capital investment may have played a role in productivity downturn. However, they did not agree about claimed estimates of the marginal productivity of public capital and the extent to which public infrastructure investments lead to increases in economic development. One group of participants saw a strong link between infrastructure investment and economic and social well-being and wanted more investment. The other group wanted to see a more efficient use of existing infrastructure based on engineering, pricing, and financing.

One of the economists, David Aschauer, had presented high-profile papers to make a case for the importance of infrastructure to the quality of life, the environment, and productivity. His work seemed to reinforce the construction industry's case for more investment in infrastructure. He cited wastewater and solid wastes as examples of infrastructure's link to the economy and to public health, and he noted how inadequate public transport was a barrier to employment for those without cars and how congestion hurts the country. Aschauer's model estimated a production function with empirical evidence that the marginal productivity of public capital was much higher than for private capital. Other participants thought infrastructure was linked to the economy, but they disputed Aschauer's conclusions and methodology.

One participant in the conference, Alicia Munnell, thought that public capital has a significant and positive impact on output at the state level, and that water and sewer systems had the largest impact or output, followed by highways, with other public capital having very small impacts. She argued that, on balance, public capital investment stimulates private investment rather than displacing it, and that the results of studies indicate that public capital has a positive impact on private sector output, investment, and employment.

Another participant, George Peterson, thought that infrastructure undersupply is as much a problem of politics as of economics. He argued that the decline in infrastructure investment does not in itself indicate that it is undersupplied and that more information is required to determine whether there is a shortfall in public capital. He saw evidence of an undersupply of infrastructure and thought that as long as benefits spill over to users outside the local taxing district, local taxpayers will provide a suboptimal level of infrastructure capital. The problem could be solved with user fees or intergovernmental matching grants. Peterson suggested that the fear of rejection by public officials leads to attempts to garner large majorities to minimize chances of rejection. However, he thought that creating authorities to invest in infrastructure without referendums is a bad idea, and he advocated the formation of business and consumer alliances to take the case to the public.

Still another participant, Joel Tarr, a historian, explained how both public and private capital spending exhibit irregular cycles of spending followed by retrenchment and stability and that spending has shifted over time among levels of government and between private and public providers. The participants concluded that privatization's advantages are not entirely clear.

The participants in the conference thought that an efficient infrastructure policy will regulate demand and investment guidelines to create an optimal policy. Some would impose costs on users that took into account the details of pavement damage and the congestion caused by different types of vehicles. Others saw the merits of the efficient pricing and investment argument but thought that political problems would create barriers. For example, current fuel tax policy indirectly encouraged shippers to use the least number of axles and the most weight per axle, thus creating the most pavement damage per haul.

In conclusion, the conferees agreed that infrastructure is important for the environment, the quality of life, and the economy and that the nation did cut back on investment in it. The question of incentives divided them, in that some wanted different incentives and others wanted more investment.

Clearly, infrastructure issues must be viewed in the context of economic-social-physical interdependencies in arenas such as economic growth, jobs, and urban problems, including housing, transportation, and public services.

To put the so-called infrastructure crisis in perspective, Fig. 4-5 shows how, as the nation entered the twentieth century, population and development ratcheted infrastructure needs upward. The main driving force was the automobile, but the increasing population and rising living standards also increased demands for infrastructure. Reliance on government increased during the 1930s Great Depression and up to the 1970s. This 40-year era included the Great Depression, World War II, the postwar recovery, and President Lyndon Johnson's Great Society.

At the time of Ronald Reagan's election in 1980, reliance on government solutions declined and the infrastructure crisis emerged. The demand for infrastructure continued to increase, so a gap opened between expectations and the government's ability to provide for needs. As a result, today's increased interest in private sector solutions is driven by the demand for infrastructure and the limits on government's supply of it. While there is substantial interest in private sector solutions, the stimulus program that Congress passed to defuse the financial crisis seemed to increase government involvement in many areas of the economy, including infrastructure.

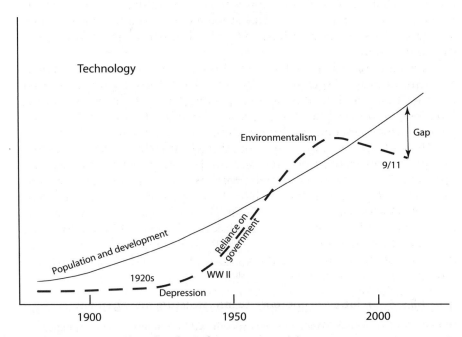

FIGURE 4-5. Timeline for the infrastructure crisis

Beginning in the late 1990s, the American Society of Civil Engineers (2007) began issuing an infrastructure "report card." The society's leaders got the idea from a report of the National Council on Public Works Improvement (1988) that included a report card. The report card attracts attention to the need for more investment, and it is often cited in the media rush that accompanies an infrastructure failure.

Infrastructure Planning

Although politics largely determines infrastructure investments, people agree conceptually that a rational approach to planning should occur within a life cycle perspective. Figure 4-6 illustrates how in this life cycle view, planning is followed by construction, operation, and renewal. An approach like this should have incentives so systems can be managed over a lifetime and not deteriorate to the point where they must be rebuilt at high cost.

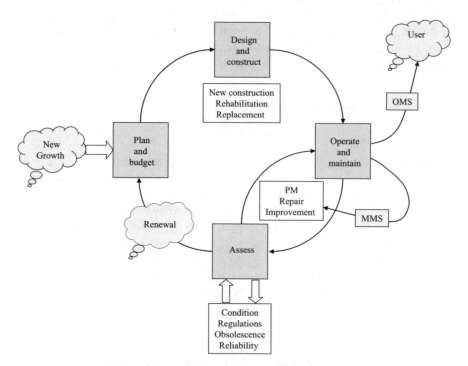

FIGURE 4-6.   A life cycle approach to infrastructure management

Note: OMS is operations management system, MMS is maintenance management system, and PM is preventive maintenance.

Capital Improvement Planning

Infrastructure planning occurs within the process of a capital improvement program (CIP), which was shown earlier in the book, in Fig. 3-4, the hierarchy of planning. The CIP offers an organizing platform to merge city planning, facility needs, and other aspects of urban growth into an integrated package. It requires a series of steps.

The first step is integrated planning, where the big picture of an urban area can be viewed. For example, a city's comprehensive master plan is a way to hang different parts of a plan on one framework to get an integrated view. Next, responsibilities to plan and develop capital programs for sectors are divided up. A sector can be an area (such as part of a city) or a function (such as transportation or water). Sectoral planning might involve an integrated plan for an area or a single-function or multifunction plan for facilities.

Using these plans, the next step is to derive the broad outlines of the capital improvements that are required—for example, a road extension or widening, new roads, a rail line, a new airport, or new water or wastewater facilities. After the broad outlines are set, groups of projects or systems can then be segregated for further planning. These can be divided into projects and subprojects or incremental project stages. Preliminary planning for these subprojects leads to costs and other economic information, and the subprojects can be programmed for their schedules of construction and implementation.

Infrastructure Condition Curves

The condition curve for infrastructure facilities offers us a way to illustrate the need for life cycle management. As shown in Fig. 4-7, a constructed facility begins in "like new" condition. This assumes that its construction is of high quality. If not, then there is an initial loss of condition due to a failure in quality control, and the condition curve can deteriorate rapidly.

If the new facility is well built, it deteriorates slowly and gives service like a new facility should. However, after it begins to deteriorate more rapidly, decline can occur suddenly, leading to a loss of functionality. The trick is to renew the facility before this starts to occur. Figure 4-7 is based on pavement condition curves, which illustrate the build-use-renew cycle. These are widely used to explain the need for the periodic renewal of pavements. Another feature of Fig. 4-7 is to show depreciation curves, which are discussed in Chapter 10.

Infrastructure Performance Assessment

Infrastructure planning is closely linked to performance assessment. Most performance indicators are associated with sectors such as transportation or

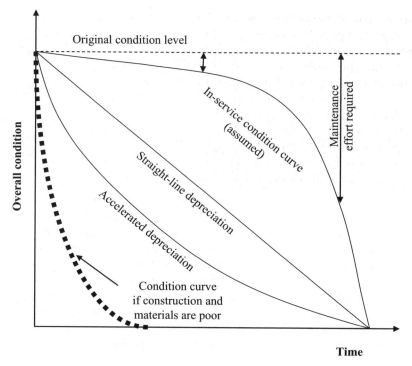

FIGURE 4-7. Infrastructure condition curves

water supply, but some measures apply to overall assessments. These will usually be associated with investment or the other high-level concerns of agencies such as the U.S. Government Accountability Office.

The National Research Council (1995) conducted a study on generalized infrastructure performance measures and adopted three categories of indicators: effectiveness, reliability, and cost. These are logical from the viewpoint of budget analysts, but they must be applied in specific ways to the categories of infrastructure. Otherwise, the metrics will not be logical.

References

American Society of Civil Engineers (ASCE). (2007). Long commutes, dirty water, delayed flights, failing dams. http://www.asce.org/reportcard/ 2005/index.cfm. Accessed August 18, 2007.

Choate, P., and Walter, S. (1981). *America in ruins: Beyond the public works pork barrel.* Council of State Planning Agencies, Washington, DC.

Department of Homeland Security (DHS). (2006). *National infrastructure protection plan*. U.S. Government Printing Office, Washington, DC.

Munnell, A., ed. (1990). Is there a shortfall in public capital investment? Conference Series 34. Federal Reserve Bank of Boston.

National Council on Public Works Improvement. (1988). *Fragile foundations: A report on America's public works*. National Council on Public Works Improvement, Washington, DC.

National Research Council. (1995). *Measuring and improving infrastructure performance*. NAE Press, Washington, DC.

National Science Foundation. (1993). *Civil infrastructure systems program guide*. National Science Foundation, Washington, DC.

5

Transportation Economics

Transportation Systems and Economic Advancement

When identifying public investments with high economic impact, people usually mention transportation systems first because they make up the largest sector among public infrastructure systems and have the most influence on economic development. Transportation systems have spurred economic development for many centuries by enabling the movement of people, goods, and information across trade routes. Examples include trade from China along the Silk Road, the expansion of the Ottoman Empire, the spreading of culture by sea and land routes, and the movement of settlers on the Oregon Trail or rail routes. Today, global trade depends on modern transportation systems that have increased our mobility greatly during the last century.

Although the Internet has ushered in much greater flows of information and telecommuting, the demand for transportation for people and goods shows no sign of abating. The demand for trips in urban areas, for air travel, and for shipping of goods continues to increase. Evidence is seen in great increases in air and road congestion, along with a big leap in the business of logistics to support Internet shopping.

On the supply side, systems to transport information have increased greatly in capacity and sophistication, but traditional transportation systems— road, air, rail, and water—show signs of strain. Indicators of strain are traffic congestion, high infrastructure cost, energy consumption, security, and public health problems from crowding. The condition of transportation systems remains a large concern across the nation.

The future of transportation is exciting to imagine, if society can overcome the barriers to system development. New sources of energy and vehicle designs (such as hydrogen and hybrid vehicles), the use of information technology, and an emphasis on bicycle and pedestrian transportation are promising trends.

This chapter presents an overview of transportation economics. It describes the modes and industries of the transportation sector, and it explains the principal economic issues that confront them.

Transportation Systems

Transportation infrastructure comprises a set of systems that provide mobility for the nation's economy. There is a lot of demand for transportation, and the supply system involves both the government and private sectors in a social contract whereby the government provides roads, airports, and other types of infrastructure and the private sector provides automobiles, airlines, fuel, and other support. This supply-demand relationship involves a number of national policy and security issues, ranging from government fuel economy standards to airport security systems.

Transportation infrastructure can be classified into road, air, mass transit, rail, and water transportation subsectors for the different modes. Each of these has its own industry and unique facilities. Examples of facilities for various modes are:

■ *roads and highways:* all rural and urban highways, roads, and streets;
■ *air:* all airports, airways, and the associated infrastructure;
■ *mass transit:* all intracity bus and rail lines;
■ *rail:* intercity passenger and freight rail lines;
■ *water:* rivers and waterways, maritime shipping, and ports and harbors;
■ *pipeline:* pipelines to transport liquids and slurries;
■ *bicycle, pedestrian:* bicycle lanes and trails, sidewalks, and paths; and
■ *intermodal:* terminals to facilitate transfer between modes.

As Fig. 5-1 shows, these modes provide choices for intercity and intracity transportation. Through these choices, travelers can use the most efficient modes to plan their schedules and routes of travel.

Although transportation can be viewed as a unified system, the management and finance of its modal sectors vary considerably. Management involves a combination of government assistance and regulation. The methods to finance roadways, transit, rail, and air travel evolved with the technologies and demands on the services and are based on each mode's unique characteristics.

At the national level, transportation policy involves several federal agencies. At the state and local levels, most responsibility is focused in single

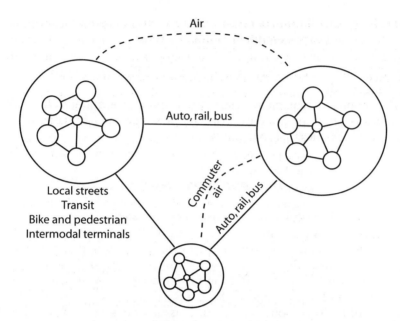

FIGURE 5-1. Intercity and intracity transportation networks

departments. Table 5-1, which presents a list of agency responsibilities, illustrates how transportation roles are distributed among various government departments (U.S. Department of Transportation 2008).

A measure of productivity in transportation would take into account the supply, cost, and performance of the different modes. This involves complex systems of statistics, which are maintained by the U.S. Department of Transportation and its subsidiary agencies, such as the Federal Highway Administration (FHWA). Though it is not possible to compute meaningful overall measures of transportation productivity across modes, you can compute productivity measures for highways, transit systems, and specific systems where data on performance and cost are available.

Transportation involves many social issues that attract policymakers. Examples include access by low-income populations to transit and travel services, the cost of transportation, public health and safety, and evacuation during emergencies. For these reasons, it is expected that government will continue to have a heavy involvement in the transportation industry but not become a major service provider.

The balance points between government and the market have been set over the years by policy decisions about matters such as public support for highway construction, regulation of the travel industries, and government incentives for transit, and congestion mitigation. Although there is no serious

TABLE 5-1. Distribution of responsibility among government agencies for various modes of infrastructure

Mode	Responsibilities
Highways and streets	The Federal Highway Administration and National Highway Traffic Safety Administration are the main federal agencies. Each state has a department of transportation or equivalent agency. Local cities and counties have street and road departments. Bicycles and pedestrian transportation are included with roads and streets.
Air	Airports are mostly owned by local governments. The Federal Aviation Administration regulates them and airline operation.
Mass transit	Mass transit is mostly a local service. The Federal Transit Administration provides assistance to systems.
Rail	Railroads in the United States are mostly privately owned, but with federal oversight and regulation. The Federal Rail Administration is the main regulatory agency.
Water	Water transportation is regulated by the Maritime Administration for ocean shipping and by the Army Corps of Engineers for the Intracoastal Waterway system. The Coast Guard has a role in water transportation.
Pipeline	Some commodities can be shipped by pipeline, and the regulatory agency is the Pipeline and Hazardous Materials Safety Administration.
Intermodal	Intermodal facilities are the general responsibility of local governments and the state departments of transportation and federal Department of Transportation.

thought that government would take over private companies such as airlines, new thinking is being done about privatizing transportation facilities.

Road Transportation Systems

Characteristics of Road and Highway Systems

Road transportation is a driver of economic development and the expansion of trade. Road networks developed historically to connect one settlement to another. To accommodate the automobile, old roads were improved and eventually became highways. As this occurred, the federal government saw an increasing need for national roads and in 1918 created the Bureau of Public Roads (BPR) to guide design and construction. The BPR was initially part of the Department of Agriculture. The FHWA is the successor organization to the BPR. The FHWA dates to 1970, after the BPR was absorbed into the U.S. Department of Transportation, which was formed in 1967. During the 1950s, the Interstate Highway System was launched, leading to today's mixed network of intercity roads. At the same time, continuing urbanization has created more demand for local and regional roads and streets (FHWA 2008).

Today, the United States has some 4 million miles of roads and local streets. This statistic is listed, along with many others, in the FHWA's (2007a) "Highway Statistics." For perspective, this compares to about 250,000 miles of roads in Great Britain.

Roads are classified by ownership and responsibility (federal, state, local, toll authority) and by type (interstate, primary highway, rural roads, principal arterial, minor arterial, collector, and local streets). Road construction is either asphalt or concrete, with some 90% of roads made of asphalt. Some gravel roads remain unpaved.

The federal aid system carries most of the nation's traffic. It includes some 47,000 miles of Interstate Highways and greater lengths of primary, urban and secondary highways. Most roads are rural and are under local control. The system that does not receive federal aid carries some 20% of the traffic, but these are often of life-or-death importance to local residents.

Road mileage in the United States is overwhelmingly rural and locally owned—77% of mileage is in rural areas. Within the road system, the National Bridge inventory shows some 596,800 bridges as of 2005 for all ownership categories. More than 96% of these are owned by state and local government agencies (FHWA 2007b).

The capacity of a roadway measures its ability to move traffic and support mobility. Levels of service are determined by design characteristics and by road condition. Design characteristics include capacity, vertical and horizontal alignment, and cross section. The guidebook for capacity is the *Highway Capacity Manual*, from the Transportation Research Board (2000). Capacity is reported as service levels, which are given as A through F: A = free flow, B = stable flow, C = stable flow (more restrictions), D = approach to unstable flow, E = volumes near capacity, and F = forced flow.

Roads and streets are built to standards that determine the cost and level of performance. In addition to their transportation functions, roads and streets provide organizing space for utility corridors and the social and economic activity of communities. They consume large quantities of raw materials and construction effort, and they are a significant source of air and water pollution. Providing mobility and reducing congestion on them is a principal issue facing many cities.

Federal oversight of roadways falls under the FHWA. State transportation departments have responsibility for the construction and maintenance of federal-aided and state roads. Local roads are administered by city and county governments. Also, authorities such as the New Jersey Turnpike Authority administer roads within their jurisdictions. The American Association of State Highway and Transportation Officials (AASHTO) issues standards, guidelines and publications. The Transportation Research Board, under the National Academy of Sciences, administers the National Cooperative Highway Research Program and issues the *Highway Capacity Manual*.

Maintaining the condition and performance of the nation's network of roads and bridges is a continuing challenge. In 1984 a task force of the Transportation Research Board (1984) developed a plan for strategic research in transportation. It identified a number of issues that led to improvements in practices for using asphalt, long-term pavement performance, the cost-effectiveness of maintenance, protecting bridge decks, improving the use of cement and concrete, and controlling snow and ice chemically. With the improved technologies, concern has shifted to finance as the most serious issue to implement the improvements and maintain infrastructure assets.

Safety is a continuing concern as well. For example, a 1967 failure of a suspension bridge in West Virginia led to a national program of bridge inspection and replacement. In 1983, almost half the nation's bridges were found to be obsolete or structurally deficient, but that proportion has since been reduced to about 28% (FHWA 2005). In 2007, a massive bridge failure occurred on the Mississippi River in Minneapolis. These failures point to the close relationship between condition assessment, investment, performance, and safety.

Traffic congestion is a growing concern that involves difficult choices. Analysts figure that the cost of lost time and fuel due to congestion is about $200 billion annually, or 2% of gross domestic product. The Texas Transportation Institute issues an annual analysis showing losses in congested areas on the order of $1,000 per peak traveler per year based on lost time and wasted fuel. This estimate has increased greatly in the last 20 years, based on constant dollars. Congestion is even implicated in global warming, due to the excessive fuel burned during delays (McKinnon 2007).

Investment Needs

Road and highway investment needs are reported in the FHWA's "Conditions and Performance Report," which estimates the required investment scenarios by all levels of government to maintain year 2002 condition and performance indicators through 2022 (FHWA 2007c). This investment would prevent average highway user costs (travel time, vehicle operating costs, and crash costs) from increasing and cover pavement and bridge preservation as well as required system expansion. The scenarios take into account the deployment of new operational technologies, including intelligent transportation systems. They do not take into account congestion pricing.

In its report, the FHWA has two scenarios, one to maintain and the other to improve the status quo. The required annual investment for 2003–22 for the "cost to maintain highways and bridges" scenario is estimated at $73.8 billion, and for the "cost to improve" scenario it is $118.9 billion. This latter estimate includes the cost to all levels of government for all highway and bridge improvements that pass a benefit-cost test. It would address the existing backlog of highway ($398 billion) and bridge ($63 billion) deficiencies and new deficiencies that arise, when they pass the benefit-cost test.

A 2007 report by AASHTO said that federal funding should rise more than 80% just to keep up with inflation. The main driver of inflation has been material costs, with labor shortages also being a big factor. This funding level would require an increase of 10 cents per gallon in the gasoline tax by 2015 (Ichniowski 2007). Table 5-2 presents the FHWA's classification of investment needs for highways and bridges. System preservation improvements make up 46.9% of the maximum economic investment (improve) scenario. This includes all capital investment aimed at preserving the existing pavement and bridge infrastructure, such as resurfacing, rehabilitation, and reconstruction. This does not include the costs of routine maintenance. Investment requirements for system expansion make up 44.5% of the maximum economic investment scenario. The remaining 8.6% is not directly modeled; this represents the current share of capital spending on system enhancements such as safety, traffic control, and environmental investments. System enhancements include items such as safety, traffic control, and environmental investments.

The road data are generated from the FHWA's Highway Economic Requirements System, which models user, agency, and societal costs for travel time, vehicle operating, safety, capital, maintenance, and emissions costs. Bridge

TABLE 5-2. The Federal Highway Administration's analysis of investment needs for highways and bridges under two scenarios

Investment need	Cost to maintain scenario	Cost to improve scenario
Preservation	$40.0 (54.1%)	$55.7 (46.9%)
Expansion	$27.5 (37.2%)	$52.9 (44.5%)
Enhancement	$6.4 (8.6%)	$10.2 (8.6%)

Note: Dollar values represent average annual costs in billions. Percentages indicate the share as part of the scenario's total cost.

Source: FHWA 2007c.

rehabilitation and replacement costs are from the National Bridge Investment Analysis System, which uses benefit-cost analysis.

Highway Performance

In the United States, automobiles provide a high degree of personal mobility. Most daily trips use personal vehicles, and highways carry the lion's share of freight in the United States. Detailed data on travel are provided by the *Conditions and Performance Report* (FHWA 2007c). Table 5-3 shows distribution of some 2.9 trillion vehicle miles traveled in 2002 in the United States. Highway mileage is mostly rural, but some 60% of travel was in urban areas in 2002. Now, rural travel is growing slightly faster than urban travel (it was growing at 2.4% in 2000 and at 2.8% in 2002). From 1982 to 1993, urban travel rates grew faster. This seems to show a reversal of urbanization trends.

TABLE 5-3. Highway miles and vehicle miles traveled by area and type of road, 2002

Functional system	Miles	Lane miles	Vehicle miles traveled
Rural areas			
Interstate	0.8	1.6	9.8
Other principal arterials	2.5	3.1	9.0
Minor arterial	3.5	3.5	6.2
Major collector	10.8	10.4	7.5
Minor collector	6.8	6.5	2.2
Local	52.9	50.6	4.9
Subtotal for rural areas	77.3	75.7	39.4
Urban areas			
Interstate	0.3	0.9	14.3
Other freeway and expressway	0.2	0.5	6.6
Other principal arterial	1.3	2.3	14.3
Minor arterial	2.3	2.8	11.9
Collector	2.3	2.3	5.0
Local	16.2	15.5	8.4
Subtotal for urban areas	22.7	24.3	60.6
Total	100.0	100.0	100.0

Source: FHWA 2007c.

The overall performance of highways can be measured by mobility and congestion, which are two sides of the same coin. If congestion increases, mobility decreases. Congestion causes added travel time, delays, and greater emissions. Vehicle miles traveled (VMT) grew faster between 2000 and 2002 than lane miles (2.5% versus 0.2% per year). Truck VMT is growing faster than passenger vehicle VMT. The indicator "percent of travel under congested conditions," or the portion of urban traffic that moves at less than free-flow speeds, increased from 21.1% in 1987 to 30.4% in 2002. The "rush hour" increased from 5.4 to 6.6 hours per day (FHWA 2007c).

Sources of Finance

Roads, streets, and bridges are mostly financed from dedicated funds at the level of government responsible. When state and local government expenditures on highways are added to those of the federal government, they came to $135.9 billion in 2002 (an increase of 33.3% in current dollars and 18.4% in constant dollars from 1997). The federal government funded $32.8 billion (24.1%), including amounts transferred to state and local governments. States funded $69.0 billion (50.8%), and local governments funded $34.1 billion (25.1%). Of the expenditures, 50.2% went for capital outlay.

Of funds for construction and maintenance, the federal government provides 28%, state governments 50%, and local governments 22%. The state portion is 34% from fuel taxes, 30% from federal grants, and 17% from motor carrier taxes. Bond issues pay for only 4% of state contributions.

Toll roads normally operate as independent enterprises, with revenues dedicated to their capital construction, operation, and maintenance.

The federal aid system utilizes the Highway Trust Fund (HTF) to collect revenues, and most state governments use a gasoline tax as well. Most local governments use property taxes and various types of fees. The HTF is the federal mechanism to collect gas tax revenues. It was created by the Highway Revenue Act of 1956 to finance the Interstate Highway System and the federally aided highway system. A mass transit account was created in 1983 to divert some of the funds to transit. The HTF receives excise tax revenues from sales of fuel, truck tires, trucks and trailers, and heavy vehicle use. Most of the HTF taxes are paid to the Internal Revenue Service by the producer, retailer, or heavy vehicle owner.

Federal funding is authorized through a series of transportation equity acts (TEAs), which occur every five years (Fig. 5-2). The first TEA bill was the 1991 Intermodal Surface Transportation Efficiency Act (ISTEA), which authorized some $155 billion over six years, of which $120.8 billion was directed toward highways. Before this, the completion of the Interstate Highway System was the focus. ISTEA created an intermodal framework for

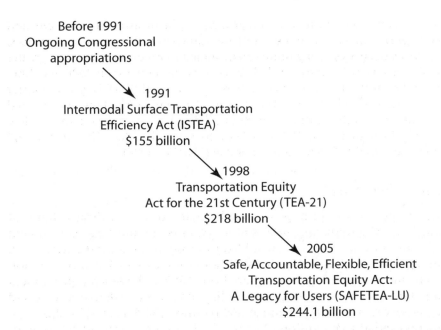

Before 1991
Ongoing Congressional
appropriations

1991
Intermodal Surface Transportation
Efficiency Act (ISTEA)
$155 billion

1998
Transportation Equity
Act for the 21st Century (TEA-21)
$218 billion

2005
Safe, Accountable, Flexible, Efficient
Transportation Equity Act:
A Legacy for Users (SAFETEA-LU)
$244.1 billion

FIGURE 5-2. The evolution of transportation equity acts and highway finance

transportation policy and emphasized state and local roles in planning. The funding actually provided by ISTEA reached nearly 100% for highways but only to about 75% for the transit funding target.

TEA-21, which covered the period from 1998 through 2003, was the largest public works bill in history and authorized nearly $218 billion for highway construction and maintenance, highway safety, and transit programs. It created a firewall to guarantee minimum funding levels of about $198 billion. It also had a "minimum guarantee" provision that each state must receive at least 90.5% of its HTF contributions. Also, funding for highways and highway safety was to be linked to HTF receipts. It retained the framework of ISTEA and addressed funding equity between the states (Northeast Midwest Institute 2007; New York State Department of Transportation 2007).

Next came the Safe, Accountable, Flexible, Efficient Transportation Equity Act: A Legacy for Users (SAFETEA-LU). It was signed in 2005 to provide funds for 2005–9. Its funding level was $244.1 billion for highways and transit. SAFETEA provided provisions for innovative finance to make it more attractive for the private sector to participate in highway projects, as well as private activity bonds, flexibility to use tolling, and broader loan policies. The bill also promoted congestion relief by giving flexibility to use road pricing to manage congestion and promote real-time traffic management. Efficiency was to be promoted through longer-lasting highways and faster construction, and provisions for environmental stewardship and greater safety for non-motorized transportation were also included (FHWA 2007d).

Mass Transit

Mass Transit Services

Mass transit offers a transportation alternative for mobility in urban areas. Most transit operations can be classified as bus or rail, although there are variations such as surface rail and underground rail, light rail, and heavy rail. The new bus rapid transit technologies are a hybrid approach, with lower-cost bus technology on a fixed route system like a rail system.

Transit is like a utility in the sense that you should be able to rely on user charges. However, alternatives for transportation are available, and transit is not always the mode of choice. Transit is efficient from a collective standpoint, but it is not always the most convenient individual choice. Consequently, transit usually relies partially on subsidies to promote its social and environmental purposes.

The use of transit has fluctuated with economic and technological conditions. It operated with horse-drawn cars even before electric power and the internal combustion engine were developed. Before 1940 many private transit companies operated, but with increased automobile travel and urbanization after World War II, ridership declined and many private companies failed. Ridership peaked in 1946 at 23.4 billion trips on trains, buses, and trolleys, and then it declined due to the low cost of fuel and resulting urban sprawl. Ridership dropped to 6.5 billion trips in 1972, then increased gradually to 9.7 billion trips in 2001 (American Public Transit Association 2005).

Federal intervention started in 1963 with a small program to enable local governments to take over ailing companies. The federal government became more active through the Federal Transit Administration (formerly the Urban Mass Transportation Administration) and provided grants to finance new systems, acquire equipment, and subsidize operating deficits. The funding varies from year to year with the federal budget.

In some cities, transit does well. In others, it is difficult to implement. Transit has advantages when a city has captive constituencies without many cars, but it faces obstacles of acceptance, finance, and operations. Bus systems are dependent on labor for operations. Transit faces security issues, which in recent years have been called into stark relief by the Madrid and London bombings and the Tokyo subway nerve gas attack. The growth of heavy rail in New York saw a decline after the September 11, 2001, terrorist attacks that destroyed parts of the subway system.

Nevertheless, the prospects for mass transit seem to be improving, even in sprawling cities like Los Angeles and Denver. In recent years, Denver's Regional Transportation District has had success with a limited light rail system, which attracts high ridership on some lines but has greatly exceeded its projected costs. Now, however, Denver is considering some heavy rail lines, including one to Denver International Airport. These can escape union clauses in the authority's contract and be built and operated in public-private partnerships that are prohibited in the union environment (Leib 2007).

Transit Investment Needs

Transit investment needs have also been reported by the FWHA (2007c). Transit investment estimates are by the Transit Economic Requirements Model, which analyzes benefits and costs to replace and rehabilitate assets, improve operating performance, and expand transit systems to serve growth in demand.

For the period 2005–24, average annual investments required are $15.8 billion for the "maintain" scenario and $21.8 billion for the "improve" scenario. Some 87% of investment required under the maintain scenario is for large urban areas of over 1 million in population. This occurs because some 92% of passenger miles are in these areas.

Air Travel and Airports

The air transportation system is a partnership between the local governments that operate airports, the airline companies that provide the actual travel vehicles, and the federal government, which handles air traffic control and air security. In the current system, airliners and privately owned aircraft travel from one publicly owned airport to another. Thus the air transportation infrastructure is mostly public, and the users are mostly private.

The airline industry goes through continuing transitions. In the United States, a highly regulated industry was deregulated in the 1980s, causing shakeouts, mergers, failures, and other changes in services. Now, airlines are a commodity industry that sells seats. Competition and deregulation have lowered the cost of a seat-mile, but low-margin airlines are sensitive to fuel costs, labor disruptions, and setbacks due to security incidents or the weather. As a result, a number of bankruptcies and mergers have occurred. In some countries, the main airline was owned by the government (such as British Airways or Air France) but has been privatized.

Infrastructure development for airport expansion, equipment, facilities, and operational support is a continuing problem due to the growth of air travel. The United States has some 15,000 landing places, but only about

20% of them are publicly owned, open for general use, and equipped with at least one paved and lighted runway. A total of 90% of the nation's passenger traffic comes through 66 airports, or some 2% of the total. Airports in the United States are mostly owned by local governments with revenues from a variety of sources such as ticket taxes, landing fees, concessions, and other use charges.

The economic implications of airport development are important. Hub airline service is critical to the growth and development of major cities. Examples can be seen from the growth of Atlanta, Chicago, Denver, Dallas, and Los Angeles. In the 1980s the development of Denver's new international airport was the state's highest economic development priority. Now, it has more than 10 years of operating experience and continues to expand to serve a growing economy in the region with trade and tourism. The airport is about to embark on a seven-year expansion and modernization program costing $1.2 billion. This expansion will comprise new gates, a new commuter terminal, expanded parking, and a terminal station for a planned rail line from Denver's Union Station. The airport has managed to keep charges per passenger down by spreading its fixed costs over a large passenger base. It is now the 5th busiest in the United States and 10th busiest worldwide (*Denver Post* 2007).

To serve the nation's airlines, the air traffic control system requires continuous upgrading to keep pace with the growth in traffic. Airport congestion is a major problem of delayed passengers, wasted fuel, and lost crew time. In the 1980s the hijacking problem was the focus of security, but after 9/11 the risks shifted to bombs and planes used to attack critical facilities.

Today's Federal Aviation Administration regulates and encourages civil aviation and promotes safety, operates the system of air traffic control for civil and military aircraft, conducts research, and regulates commercial space transportation. It operates airport towers, air traffic control centers, and flight service stations. It develops air traffic rules, assigns the use of airspace, and controls air traffic. It also builds or installs visual and electronic aids to air navigation, including communications, radar, computer systems, and visual display equipment. After the terrorist attacks of 9/11, Congress created the Transportation Security Administration to take over civil aviation security from the Federal Aviation Administration. No one is exactly sure how future paradigms for air travel will unfold, whether through today's "hub-and-spokes" system or new approaches with more travel between small hubs with smaller aircraft.

Intercity Rail and Bus Systems

Bus and rail compete with air travel for passengers traveling between cities. Air travel reduced demand for these slower modes, but they remain viable in some market niches and freight rail continues to serve important markets. Rail systems for passengers and freight in the United States are operated mostly by the private sector, but the government regulates and subsidizes them to some extent. Today, rail systems carry only a tiny fraction of passenger traffic and a minority of freight loads.

Intercity Rail

Rail service in the United States began with the first steam-powered train in 1830. By the Civil War, 30,000 miles of track were in service. The government subsidized railroads through land grants, but subsidies were terminated in 1870. The first transcontinental railroad was completed in 1869. In 1900, railroads were growing in number and their stocks were hot on Wall Street. By 1910, trains carried 95% of all intercity traffic, but soon this percentage began to fall. During World War I, the federal government nationalized the trains. By 1929, intercity train traffic had fallen by 18% from its peak in 1920. During the 1930s, new technologies such as the diesel engine, higher speeds, aerodynamic design, and air conditioning maintained the popularity of trains. By 1939, passenger rail travel had increased 38% in six years, but the number of passengers was still less than half the 1920 numbers. During World War II, U.S. railroads avoided nationalization, and they saw record increases in passengers and freight. Trains moved so many troops that the public had to delay personal travel. This set the stage for a postwar decline in train usage. During the 1950s, airlines, automobiles, and buses increased their intercity travel. They also received government subsidies to highways and airports (Boyd and Pritcher 2007).

By 1970, airlines carried 73% of passenger travel, and the railroads' share had dropped to 7.2%. The government created the National Railroad Passenger Corporation (Amtrak) in 1971 to provide balance in options and reduce automobile congestion. Amtrak did well during the 1970s oil embargo, but ridership was not maintained afterward. Today, Amtrak remains the only nationwide passenger rail operator in the United States.

Trucks and rail each carried about 40% of domestic U.S. freight in 2002 as measured in ton-miles. Water carried 9%, and air carried only 0.2%. However, when value is considered, rail drops to 3.7% and trucks' share rises to 74%, because most rail traffic is in lower-value commodities (U.S. Bureau of Transportation Statistics 2007a).

In 1976, the federal government created the Consolidated Rail Corporation (Conrail). Later, it was freed from operating commuter rail service. By 1981, Conrail did not require federal subsidies any more. In 1987, the federal government sold its interest in Conrail in the largest initial public stock offering in history to that time. This returned the system to the private sector as a for-profit corporation (Conrail 2007).

In addition to long-haul railroads, some 400 short-line railroads operate in the United States. An example is Watco Companies, which holds some 16 short-line companies. Pennsylvania is the state with the most of these, some 55. The most commonly shipped commodity is coal, at 20%. The total 2004 revenue of this industry was about $3 billion, and it employed 12,463 people in 2006. These are represented by the American Shortline and Regional Railroad Association and are important in hauling commodities such as farm and manufacturing goods and supplies (Matthews 2007).

In densely populated countries, passenger rail remains essential. In Europe, the Eurotunnel has connected England to the Continent via rail. There, freight rail can haul a much greater percentage than it does in the United States.

Developing countries see rail as a strategic investment option. China's rail sector is very active and has opened several new lines, including a long-distance train that passes over 16,000-foot elevations on its way to Tibet.

Intercity Bus Transportation

The intercity bus industry is much smaller than in the past. It is represented by the American Bus Association (2008), which reports some 3,500 motor coach businesses in the United States and Canada. These include large players, such as Greyhound Lines, and many smaller bus lines.

Greyhound Lines is the largest intercity bus company in North America, serving 2,200 destinations in the United States. It was founded in 1914, and by 1926 the company was known as Greyhound Lines and was making transcontinental trips. It suffered in the Depression, but it later recovered and by World War II had 4,750 stations and nearly 10,000 employees. After the war and with the Interstate Highway System, automobile travel increased and ridership on Greyhound and its parallel system Trailways declined. Continued competition from airlines has caused further consolidation in long-distance intercity bus transportation (Greyhound Lines 2008).

Water-Based Transportation

The water transportation system focuses on ocean shipping, ports and harbors, the Intracoastal Waterway, and navigation along the nation's river systems. International trade increases every year, and much of it is by ship. This requires a system of ports and harbors to create intermodal nodes whereby goods can pass to rail or truck systems and then be transported to other terminals for distribution.

The ports system is important for both military and economic reasons. Cities compete for port traffic as generators of business activity. Ports require infrastructure facilities for berthing of ships, loading and unloading, and the transportation of goods inland. Rail networks leading into ports handle substantial traffic. The harbor area must be maintained, and environmental protection must be assured. Most of the infrastructure management work at ports is done through public management agencies, such as port authorities.

The major ports on the Atlantic and Pacific coasts handle mostly container freight, whereas the major Gulf Coast ports handle mostly tanker and dry bulk freight. In 2000, the top 10 ports in the United States handled 58% of vessel calls. The greatest activity was at the Port of Los Angeles–Long Beach. In 2000, it handled 5,326 vessel calls for tanker, dry bulk, and cargo freight shipments. In terms of vessel calls, the next 9 largest U.S. ports were Houston, New Orleans, New York, San Francisco, Philadelphia, Hampton Roads, Charleston, Columbia River, and Savannah (U.S. Bureau of Transportation Statistics 2007b). The security at ports has become a much greater issue than in the past. Terrorists might try to slip a weapon of mass destruction past a port in a container, and some shipments, such as liquefied natural gas, are attractive targets for a terrorist attack.

The United States has about 21,000 miles of inland waterways that provide alternative shipping for large, bulk commodities. These include navigable rivers, the Intracoastal Waterway system, which provides navigation routes along the Atlantic and Gulf coasts, and the Great Lakes system. There is one state-maintained system, the 520-mile-long New York State Barge Canal System. These inland waterways are transportation arteries for mostly bulk items such as commodities, petroleum, foodstuffs, building materials, and manufactured goods (Schilling et al. 1987).

Pipeline transportation is also one of the freight categories maintained by the U.S. Bureau of Transportation Statistics, which catalogs trucking, railroad freight, inland waterways, and air freight. Pipeline transportation is mainly for petroleum, petroleum products, and natural gas. Pipeline statistics are maintained by the Office of Pipeline Safety of the U.S. Department of Transportation. On a local basis, pipelines also carry chemical products, water, slurries, and other liquids and gases used in manufacturing.

References

American Bus Association. (2008). About the ABA. http://www.buses.org/. Accessed May 15, 2008.

American Public Transit Association. (2005). Historical ridership trends. http://www.apta.com/research/stats/ridershp/ridetrnd.cfm. Accessed January 12, 2005.

Boyd, L., and Pritcher, L. (2007). Brief history of the U.S. passenger rail industry. http://scriptorium.lib.duke.edu/adaccess/rails-history.html. Accessed July 27, 2007.

Conrail. (2007). A brief history of Conrail. http://www.conrail.com/history.htm. Accessed July 27, 2007.

Denver Post. (2007). "Expansion of DIA is vital to economy." July 31.

U.S. Federal Highway Administration (FHWA). (2005). *Status of the nation's highways, bridges, and transit: 2002 conditions and performance report.* http://www.fhwa.dot.gov/policy/2002cpr/es7.htm. Accessed January 11, 2005.

———. (2007a). Highway statistics. Office of Highway Policy Information. http://www.fhwa.dot.gov/policy/ohpi/qfroad.htm. Accessed February 8, 2007.

———. (2007b). National bridge inventory. http://www.fhwa.dot.gov/bridge/owner.htm. Accessed February 8, 2007.

———. (2007c). Status of the nation's highways, bridges, and transit: 2004 conditions and performance. http://www.fhwa.dot.gov/policy/2004cpr/execsum.htm. Accessed March 18, 2007.

———. (2007d). A summary of highway provisions in SAFETEA-LU. http://www.fhwa.dot.gov/safetealu/summary.htm. Accessed March 20, 2007.

———. (2008). Highway history. http://www.fhwa.dot.gov/infrastructure/history.htm. Accessed May 15, 2008.

Greyhound Lines. (2008). About Greyhound. http://www.greyhound.com. Accessed May 15, 2008.

Ichniowski, T. (2007). "State DOTs seek federal boost for road aid." *Engineering News-Record,* 258(11), 9.

Leib, J. (2007). "Federal project boosts RTD." *Denver Post,* July 31.

Matthews, R. (2007). "Railroads seek tax aid." *Wall Street Journal,* January 20–21.

McKinnon, J. (2007). "Bush plays traffic cop in budget request." *Wall Street Journal,* February 5.

New York State Department of Transportation. (2007). What is Tea-21? https://www.nysdot.gov/portal/page/portal/programs/tea21/what-tea. Accessed March 20, 2007.

Northeast Midwest Institute. (2007). What is the Highway Trust Fund? http://www.nemw.org/HWtrustfund.htm. Accessed March 20, 2007.

Schilling, K., Copeland, C., Dixon, J., Smythe, J., Vincent, M., and Peterson, J. (1987). *The nation's public works: Report on water resources.* National Council on Public Works Improvement, Washington, DC.

Transportation Research Board. (1984). *Strategic transportation research study: Highways.* National Research Council, Washington, DC.

———. (2000). *Highway capacity manual 2000.* National Research Council, Washington, DC.

U.S. Bureau of Transportation Statistics. (2007a). Shipment characteristics by mode of transportation for 2002. http://www.bts.gov. Accessed July 27, 2007.

———. (2007b). Top 25 U.S. ports by cargo vessel type and calls, 2000. http://www.bts.gov. Accessed August 19, 2007.

U.S. Department of Transportation. (2008). About DOT. http://www.dot.gov/about_dot.html. Accessed May 15, 2008.

6

Land and Water Resources, Energy, and the Environment

The Economics of Natural Resources and the Environment

Whereas many infrastructure decisions can be based on relatively clear benefit-cost calculations, the natural resources and environmental sector is unique because of its many public interest issues. Thus, the economics of this sector must address the sustainability of resources as well as use, and this requires valuing resources for all uses, including leaving them in their natural state.

Infrastructure managers often deal with issues of the environmental sector. Public utility executives work with it in their environmental work, developers will see opportunities and obstacles in it, and government executives may regulate it. Environmental and natural resource subjects span a wide range of topics, and you can find free-standing textbooks on resource economics, agricultural economics, energy economics, environmental economics, and other categories.

This chapter addresses the economic issues of natural resources and the environment, and the key industries that depend on them: land use, energy, water, and mining. The chapter considers both the use and sustainability of natural resources. The focus on the use of renewable and non-renewable natural resources considers minerals, water, and energy resources. The other focus considers the environment as a natural system that must be sustained.

To illustrate how these two perspectives interact, Fig. 6-1 shows the natural environment as comprising renewable and nonrenewable natural

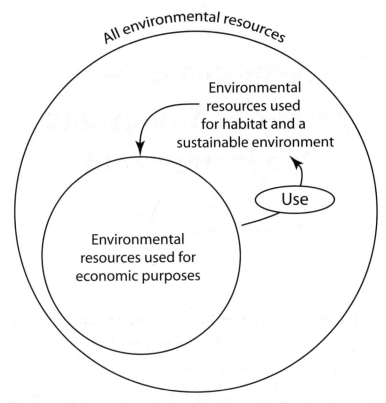

FIGURE 6-1. Environmental resources for the economy and to sustain life

resources and the habitat to sustain all life. Extracted or diverted resources are taken from the natural environment and, in some cases, returned after use. The resource returned to the environment might be about the same as the one diverted (such as water), or it might be transformed, as in the case of petroleum, which is burned and returned in the form of heat and air pollutants. Figure 6-1 could be made more complex to show other mechanisms and interactions of materials and energy budgets.

Figure 6-2 shows a framework for how the natural resources and environmental sectors interact. In the figure, resource extraction and use are shown as taking from the environment (natural systems) to be processed through infrastructure (constructed systems) to produce goods and services for society (human systems). Environmental quality and sustainability are shown as regulated issues, and the protection of public health is shown as a social issue that involves health issues affected by the natural environment.

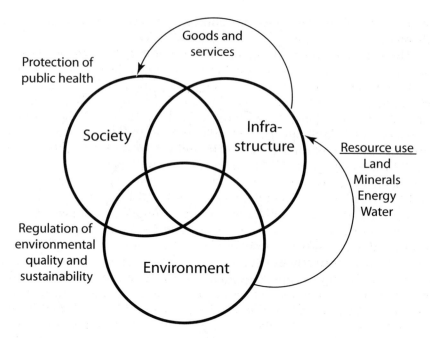

FIGURE 6-2. The environment, infrastructure, and the economy

Environmental health involves the issues of concern to environmental scientists and engineers: the pollution of land, air, and water and how it affects health. It also includes subsidiary issues, such as indoor air quality, noise pollution, and aspects of infectious disease.

Land Use

Land, capital, and labor constitute the three basic factors of production. Land use and population growth are the primary economic drivers for the interactions of the environment, infrastructure, and the economy.[1] Ownership of land often creates conflicts between owners and nonowners. The United States normally does not experience violence over land tenure, but in some countries land tenure issues involve intense conflicts over social equity. Land resources include surface and the subsurface assets. Access to mineral deposits and to oil and gas is also often regulated by government

---

[1] These subsystems (environment, infrastructure, and economy) go by different names. For example, they might be called the natural, built, and human environments.

actions. Accounting for these in a national framework was addressed in Chapter 2.

The U.S. Department of Agriculture's Economic Research Service prepares periodic surveys of land use. In its most recent inventory, the nearly 2.3 billion acres in the United States were distributed as forest use land, 651 million acres (28.8%); grassland pasture and range land, 587 million acres (25.9%); cropland, 442 million acres (19.5%); special uses (primarily parks and wildlife areas), 297 million acres (13.1%); miscellaneous other uses, 228 million acres (10.1%); and urban land, 60 million acres (2.6%) (Lubowski et al. 2007). The miscellaneous category includes rural developed land, marshes, swamps, deserts, and the like; Alaska accounts for 131 million of the 228 million acres in this category.

Land use in cities is only 2.6% of the total, and it includes the uses described in Chapter 3: residential areas; office, commercial, and government buildings; hospitals, schools and churches; industrial areas and warehouse districts; retail areas and amusement and recreation centers; public safety facilities; and transportation and utility nodes. In rural areas, land uses include rural development areas, farms, recreation and open space, resource extraction, and rural-industrial complexes. Thus, all land in the United States is some 2.3 billion acres, and developed land is some 60 million acres. This means there are some 5 persons per acre for all urban land in the United States. Another indicator is acres per capita in the nation, which would be about 7 acres per capita for the U.S. population of about 300 million. This gives the United States a population density of 31.7 persons per square kilometer, whereas for Bangladesh it is 1,040 persons per square kilometer (*Economist* 2007).

The industry structure for land use revolves around real estate and land development, which involve large sums of money. Chapter 2, which included a summary of wealth, showed that most nonfinancial national wealth is in the built environment where land is used for structures, buildings, and other similar features.

Although the supply of land is fixed, it responds to market and regulatory forces. Land uses are controlled mostly by local legislation, with an overlay of federal law for purposes such as environmental protection, national security, and other matters of public interest. Many infrastructure and environmental managers will be working in one way or another with land development.

Whether land is capital or a separate factor of production has been a long-standing economic issue and has led to the concept of land rent. Economists consider that the rent for land use differs from that for capital items, such as a house or a car. They consider land as a separate factor of production because there is only so much of it and its supply is inelastic. A related economic term is "rent-seeking," which means that individuals are looking for a profit without contributing anything to productivity. For example, if you gain control of a piece of land and you know some government action will increase its value, you are engaged in rent-seeking in this context.

If you consider land as just another asset on which to earn a rate of return, it is like capital and the distinction diminishes. This seems consistent with the accounting framework that identifies capital, labor, energy, materials, and service inputs (KLEMS) as the main factors of production (see Chapter 2). It also fits with an economic analysis that considers the cost of all inputs, capital or operations, in a rate-of-return analysis.

The iron triangle of the land development industry (Fig. 6-3) helps to explain the roles of infrastructure and environmental managers within it. This triangle has development interests at one corner, officials promoting economic development and/or land use controls at another corner, and elected officials promoting growth at a third corner. Development interests include the financiers, engineers, contractors, builders, and building suppliers with business interests in land development.

Mining, Minerals, and Materials

Minerals, petroleum, and natural gas are nonrenewable natural resources extracted from land, with important impacts on the environment and our energy future. This section discusses mining, minerals, and materials. Oil

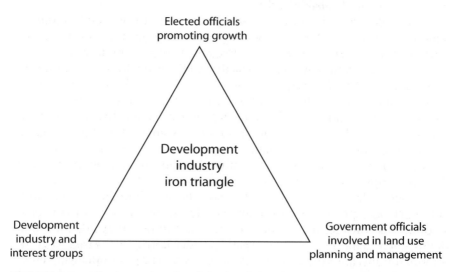

FIGURE 6-3. The iron triangle of the land development industry

and gas are discussed in the next section, which covers energy and contains a brief discussion of petrochemicals and other petroleum derivatives.

The Flow of Materials Through the Economy

Materials flow through the economy along a chain of extraction, use, and recycling/disposal. Mining and minerals are classified as metals, industrial materials, coal, and precious minerals. Extractive industries for mining and quarrying produce raw materials such as gravel, stone, and iron. These are processed into steel, cement, fertilizer, and other industrial materials. Recycling occurs to create scrap iron, aluminum, glass, and other materials. Here are examples of minerals in each of the four categories:

■ *metals:* bauxite, copper, gold and silver, iron ore, lead, mercury, molybdenum, nickel, tin, and uranium;
■ *industrial materials:* sand and gravel, sulfur, salt, limestone, phosphate, oil shale;
■ *coal:* different types of coal and fuelstocks; and
■ *precious minerals:* diamonds and other gemstones.

According to the U.S. Geological Survey, in 2004 the total value of U.S. mineral production was $45.7 billion. Most of this was industrial minerals, at $33.2 billion, whereas metals were $12.5 billion. Industrial materials include construction materials used in infrastructure. Nine commodities (other than coal) accounted for nearly 80% of production on the basis of value: crushed stone, portland cement, sand and gravel, copper, gold, iron ore, molybdenum, lime, and salt (Smith 2007).

The materials industries are close partners with the mining industry. Materials that cannot be recycled enter the waste stream for ultimate disposal. Managing them is the job of the waste management industry.

Severance taxes may be imposed on the extraction of nonrenewable resources. This provides a way for the political jurisdiction losing the resources to recoup compensation for its loss of future tax revenue. A severance tax is a tax imposed on the extraction of a nonrenewable resource such as a mineral or on a resource such as timber that might take a long time to be renewed.

One of the big issues for the mining and resource extraction industry is how to achieve and maintain its economic competitiveness while complying with rules about health, safety, and the environment. With population growth, the demand for construction materials continues to grow, and the mining of sand and gravel, copper, iron ore, molybdenum, and lime will continue to grow with it. Also, coal mining seems sure to grow, given our thirst for energy. However, all these are subject to stringent environmental rules.

Employment in Mining

Mining and resource extraction is a significant employment category. Data from the U.S. Bureau of Labor Statistics (2007) show 660,000 total jobs in mining, including oil and gas extraction. Of these jobs, 80,000 are in coal mining, 38,000 are in metal ore mining, and 118,000 are in nonmetallic mineral mining and quarrying (July 2007 data, seasonally adjusted).

Regulatory Control

Government activity in the mining industry is focused on regulation for health, safety, and the environment. Mining is regulated under the Federal Mine Safety and Health Act, which is administered by the U.S. Mine Safety and Health Administration. Mine safety is a continuing issue. In 2007, for example, a coal mine collapse at the Crandall Canyon Mine in Utah trapped 6 miners deep underground. This followed a 2005 Sago mine explosion in West Virginia that killed 12 miners (*Denver Post* 2007).

From 1910 to 1996, the United States had a Bureau of Mines in the Department of the Interior. It studied mining and minerals, encouraged health and safety, studied resource conservation, promoted economic development, and studied efficiency in the mining, metallurgical, quarrying, and other mineral industries (U.S. National Technical Information Service 2007). Some of its functions have been transferred to the U.S. Geological Survey, which publishes information on minerals.

The Iron Triangle of the Mining Industry

In the iron triangle of the mining industry, the mining companies and their support groups that sell equipment and services are at one corner. The government agencies that regulate or promote mining and minerals are at another corner, and the elected officials who are involved are at the third corner. With the downsizing of the mining industry in the United States, the extractive part of this industry is not as robust as it once was here at home, and much of its activity has shifted to other countries. However, given the nature of the demand for energy, the coal mining industry in the United States remains robust.

Energy Production and Use

Energy in the Economy

Given the rate of economic growth in China and many other countries, energy has become a hot-button issue. This section describes energy production and use in the United States. Much of the information is from the U.S. Energy Information Administration (2007).

Today's energy economy has been evolving for less than 200 years. Before the Industrial Revolution, energy use was from basic sources, including wood, peat, coal, wind, and water. Horses and draft animals provided much of the muscle that is now provided by machines. Energy powered the Industrial Revolution, beginning with steam power and requiring increased coal mining. Late in the nineteenth century, petroleum started to become important, and the electricity era took off after Thomas Edison invented the light bulb. The early twentieth century saw the development of energy and utility trusts and the rise of utility barons. In parallel, there was a good bit of labor unrest in the energy and mining industries. During the 1930s oil development took off in the Middle East, and the cheap oil era started. Nuclear power became a reality in the 1950s but was curtailed in the 1980s. The embargo by the Organization of the Petroleum Exporting Countries (OPEC) in 1973 was a wakeup call, and today limits on the global supply of petroleum seem near. Renewable energy sources are attracting attention, but they do not appear adequate to supplant oil, natural gas, and coal in the near future.

Energy Sources and Consumption by Sector

Data on energy sources and consumption by producing and consuming sectors are maintained by the U.S. Energy Information Administration (2007). Figure 6-4 shows the sources of basic energy: petroleum (39.8%), natural gas (22.4%), coal (22.6%), nuclear (8.2%), and renewables (6.8%). The percentages change slightly from year to year; Fig. 6-4 is based on 2006 estimates.

Electric power is the largest user of basic energy, at 38.9%. This is followed by transportation at 27.7%, industry at 22.1%, and residential/commercial users at 11.1%. Electric power is recycled into the other consuming sectors (see Table 6-1), so to get a total breakdown of all energy use, you would need to reallocate electricity to its consuming sectors. This would show, for example, that residential and commercial users account for much more of coal consumption than Fig. 6-4 indicates. This happens because much electricity comes from coal-fired plants. When these uses of electric power are reallocated to the sectors listed in Table 6-1, we see that for all energy use in the nation, the residential and commercial sectors account for the most, at 35.6% and 33.4%, respectively, followed by industry at 26.7% and transportation at 0.2%.

Oil for Energy and Petrochemicals

Infrastructure and environmental managers face many development and regulatory scenarios related to oil and gas. Crude oil and refined fuels such

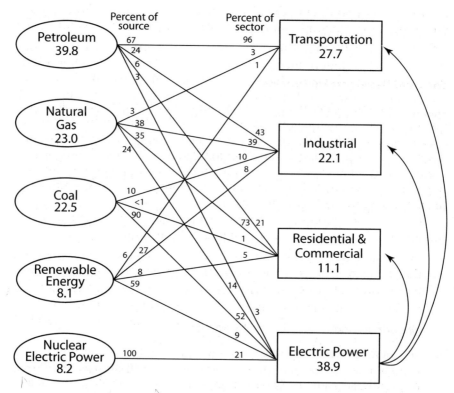

FIGURE 6-4. Production sources and consumption sectors of basic energy

Source: U.S. Energy Information Administration 2007.

TABLE 6-1. Electric power use, 2005

Sector	Megawatts per hour	%
Residential	1,359,227,107	35.6
Commercial	1,275,079,020	33.4
Industrial	1,019,156,065	26.7
Transportation	7,506,321	0.2
Direct use of self-generated power	154,700,367	4.1
Total end use	3,815,668,880	100.0

Source: U.S. Energy Information Administration 2007.

TABLE 6-2.   Petroleum statistics

U.S. crude oil production, barrels per day (bbl/day)	5,102,000
U.S. crude oil imports, bbl/day	10,118,000
U.S. petroleum consumption, bbl/day	20,687,000
U.S. dependence on net petroleum imports	58.2%
U.S. proved reserves of crude oil, 2004	20,972 million bbl

Source: U.S. Energy Information Administration 2008a.

as gasoline, diesel fuel, and heating oil are referred to as "petroleum," which also includes derivatives and petrochemical products. Natural gas is not, strictly speaking, petroleum but may be used with it as a petrochemical feedstock.

Petroleum provides about 40% of total U.S. energy needs, including 96% of transportation needs and 45% of industrial needs. The 21% of residential and commercial needs it provides come mostly from propane and related fuels. Natural gas provides another 23% of total energy needs.

The imbalance in U.S. petroleum use is shown in Table 6-2, which indicates that for the United States' total consumption of about 21 million barrels per day, the nation is almost 60% dependent on imports.[2] The table shows how the nation's supply of and demand for oil are out of balance. The country's 60% dependence on oil imports is an increase from about 35% in 1973 during the OPEC oil embargo. Obviously, energy security is a serious issue.

After the 1973–74 oil embargo, the United States established a Strategic Petroleum Reserve. As of June 2007, it contained some 690 million barrels, or a little over a month's supply at current use rates (U.S. Energy Information Administration 2007).

The oil and gas industry operates globally to extract supplies to support the demand for energy. In the United States, petroleum products are provided by private sector companies, such as Exxon, in an industry with a high degree of vertical integration that also refines and distributes these products. Some countries have state-owned companies, such as Mexico's PEMEX, which have complete control over the supply chain, from production through distribution.

The United States has experienced ups and downs in the price of gasoline. For a simple look at why prices vary, three components can be analyzed. Given the fluctuating cost of crude oil, one component of price is the

---

[2]The U.S. Energy Information Administration (2008a) presents tables to show how the dependence on imports is computed.

cost of a barrel of oil divided by 42 gallons. A second component is the gas tax, shown at $0.40 per gallon, although this will vary by state. The third cost is processing and distribution. Thus, we can take as a benchmark that for a price of $3.50 per gallon of gasoline, with oil at $100 per barrel and the gas tax at $0.40, the processing cost will be $0.72 per gallon. This figure will change, admittedly, but it gives us a rough estimate of cost. Keeping tax and this processing cost constant, we can show how gas prices will increase with the cost of oil (Table 6-3). These very approximate figures show how a combination of high crude oil prices and a demand-and-supply-induced spike in the processing cost can raise gas prices sharply.

Natural Gas Supply and Demand

Natural gas provides about 22% of the basic energy used in the United States. It competes with electricity as a fuel for heating, cooling, and cooking in buildings. It supplies 37% of the basic energy for industry, 72% of the energy for buildings, and 16% of the energy to generate electricity. The main sources of natural gas are U.S. wells and pipeline imports from Canada and Mexico. For imports by ship, it can be converted to liquefied natural gas through supercooling and later converted to gas by warming (U.S. Energy Information Administration 2007).

Natural gas is sold by volume in cubic feet or by heat content. The U.S. Energy Information Administration estimates that production in 2007 was 18,243 billion cubic feet and consumption was 21,932 billion cubic feet, leaving the nation less dependent on imports than it is for petroleum. Figure 6-5 shows how the natural gas industry works with production, transmission, and distribution facilities.

The gas industry is not very integrated. The largest production companies—such as Exxon, Texaco, Standard Oil of Indiana, and Mobil—are also involved in oil production. The largest transmission companies, which are separate from the production companies, include firms such as El Paso Natural Gas Company, Columbia Gas Transmission Corporation, Tennessee Gas Transmission Company, and the Natural Gas Pipeline Company. Distribution companies, which resemble electric and water utilities, include

TABLE 6-3. Components of the price of gasoline

Oil price ($ per barrel)	50.00	75.00	100.00
Crude oil ($)	1.19	1.79	2.38
Gas tax ($)	0.40	0.40	0.40
Processing ($)	0.72	0.72	0.72
Gas price per gallon ($)	2.31	2.91	3.50

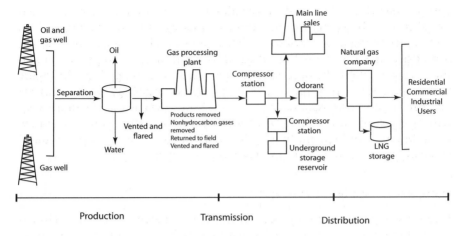

FIGURE 6-5. How the natural gas industry works

Source: U.S. Energy Information Administration 2007.

Southern California Gas Company, Pacific Gas and Electric Company, Inter-North Incorporated, and Consolidated Gas Supply Corporation. Interest groups for the natural gas industry include the American Gas Association and the Gas Research Institute, which was started in 1978.

The technology of the natural gas industry continues to evolve. In addition to new techniques for extracting gas, transmission methods have advanced. For instance, today's pipelines are much larger and operate at much higher pressure than those of only a few decades ago.

The Coal Industry

Coal is the fuel produced in greatest quantity in the United States. It is used to generate about half the nation's electric power and it is also a basic energy source in industries such as steel, cement, and paper that require large energy inputs. Coal mining has a high profile because of the cost of energy and the environmental impacts of combustion.

Coal provides 22.5% of total U.S. energy needs, and 90% of it is used as fuel for electric power production, which derives 52% of its basic energy from coal burning. Electric power production is, in turn, an important source of energy for residential, commercial, and industrial activities. Statistics show that the U.S. production of coal in 2004 was 1.1 billion short tons, or about 18% of the world total. China produces the most coal, at some 36% of the world total. The total production of coal in the United States is valued at about $26 billion, or almost as much as all industrial materials combined.

World coal production at the same sales price would be valued at some $143 billion per year (U.S. Energy Information Administration 2007).

As a fossil fuel, coal is a nonrenewable source formed as sedimentary rock that evolved from ancient plants that were converted by heat and pressure over millions of years. Coal has been used as a fuel around the world for centuries. In about 1800, it became the main energy source for the Industrial Revolution, with new railway systems being prime users. By 1900, the United States, Britain, and Germany were the main producing countries. Then oil became the fuel of choice, and direct coal use was replaced mostly by oil, natural gas, or electricity. Labor issues became important in the coal mining industry after about 1890. Now, environmental and health issues are important as well. In spite of its problems, coal remains the cheapest energy source and is the primary fuel used in electricity generation in many countries (U.S. Energy Information Administration 2008b).

The type of coal used—which can be lignite, subbituminous, bituminous, or anthracite—determines its carbon content and energy potential. Lignite is crumbly with high moisture content and, being 25% to 35% carbon, has the least energy content. It is mainly used for electric power generation. Some 21 lignite mines produce about 7% of U.S. coal supplies, mainly in Texas and North Dakota. Another 42% of U.S. coal is subbituminous, which is 35% to 45% carbon, and Wyoming is its leading source. Bituminous coal, which is 45% to 86% carbon, has two to three times the energy of lignite. It accounts for about 50% of U.S. production, and West Virginia, Kentucky, and Pennsylvania are its largest sources. Anthracite, which is some 92% to 98% carbon, provides less than 0.5% of U.S. coal. All U.S. anthracite mines are in northeastern Pennsylvania.

The United States has the largest coal reserves in the world, with some 267.6 billion short tons. This is enough to last approximately 236 years at current production levels. In 2005, U.S. coal production was 1,132 million short tons, which was an all-time record.

About two-thirds of U.S. coal production is from shallow surface mines, which produce less-expensive coal than underground mining. More than half of U.S. coal comes from the Western Coal Region, with Wyoming as the main producer. Mainly, this region has some of the largest surface mines in the world. The Appalachian Coal Region produces more than a third of U.S. coal, with West Virginia as the main producer and second nationally, after Wyoming. This region has large underground mines and small surface mines. Texas is the main producer in the Interior Coal Region, which has midsized surface mines. Western production of low-sulfur subbituminous coal surpassed Appalachian production in 1998, and interior zone production diminished.

Shipping coal can cost even more than mining it. Thus, there is an economic incentive to build coal-fired electric power plants near coal mines to lower costs. Some 68% of coal in the United States is transported by train, with the rest moved by barge, ship, truck, and slurry pipeline. Barges are the cheapest, but they cannot reach all points. Shipping coal is an important business for the deregulated railroads, which can increase their profits with longer runs and larger unit trains with high-capacity cars to ship coal to utilities.

Like other energy commodities, coal prices are variable. Prices surged around 1980 due to inflation, anticipation of rising oil prices, and the demand for electric power. This spurred coal producers to look for new reserves and to open mines. Utilities entered into long-term contracts that eventually guaranteed prices that were way above the market price. By 1985, new economic and regulatory factors—such as rising oil and natural gas prices, deregulation, increasing competition among coal producers, and international competition—changed the picture. Today's coal industry is bigger, more competitive, and more integrated than in the past. This makes it difficult for smaller players to compete.

On the supply side, coal prices are controlled by productive capacity, coal quality, geology, location, and competition among producers. On the demand side, they are influenced by competition with other fuels.

Natural gas has captured some electric generating capacity from coal-fired units, and the turbine-based combined-cycle system has made natural gas the most popular fuel for electric power generation. This has slowed the growth of new coal-fired generation capacity, but coal consumption is still increasing.

The main obstacle to more coal use is its environmental impacts, which fall into four categories: the destruction of land; the pollution of water; the production of carbon dioxide, the main greenhouse gas; and the production of sulfur, nitrogen oxides, and mercury emissions. Sulfur and oxygen form sulfur dioxide, which combines with moisture to produce acid rain. Nitrogen oxides are linked to smog and to acid rain. Mercury settles in the water and can accumulate in fish and shellfish.

The coal industry is working on these problems. It tries to clean coal before it leaves mines, and it also tries to find low-sulfur coal. Scrubbers remove sulfur from the smoke emitted by power plants. New technologies are being sought to remove sulfur and nitrogen oxides from coal or to convert coal to a gas or liquid fuel.

The mercury problem has been particularly difficult. Mercury is a naturally occurring element that exists as elemental or metallic mercury, inorganic mercury compounds, and organic mercury compounds. Metallic mercury is used in thermometers, fluorescent light bulbs, and some electrical switches. Inorganic mercury has been included in products such as fungicides, antiseptics, and disinfectants. Organic mercury compounds are formed when mercury combines with carbon. Microscopic organisms convert inorganic mercury into methylmercury, which is the most common organic mercury compound in the environment and which accumulates up the food chain.

Burning coal and using mercury in manufacturing increase the mercury in the atmosphere, lakes, and streams. The primary way people are exposed to mercury is by eating fish containing methylmercury, which accumulates up the food chain. Commercial saltwater fish, such as sharks, swordfish, and kingfish, can contain high levels of methylmercury (U.S. Environmental Protection Agency 2007).

Renewable Energy Sources

Renewable energy sources include hydroelectric, wind, solar, geothermal, biomass, and ethanol. They provide some 6.8% of U.S. needs. Table 6-4 shows the distribution among the categories of renewable sources. This use of renewables includes all applications, such as direct heating using biomass in the form of wood. For electricity generation, the uses are distributed as shown in Table 6-5. Although hydroelectricity generation in the United States is 74% of generation by all renewable sources, it is only $0.74(6.8\%) = 5.0\%$ of total basic energy production.

Nuclear Energy

Nuclear electric power currently provides some 8.2% of U.S. electric power production, and its used might grow in the future, given the environmental problems with coal as a basic fuel. According to the U.S. Energy Information Administration (2007), there are 104 operable nuclear reactors in the United States and 443 in the world. The largest U.S. plant is at Palos Verde, California, with a capacity of 3,733 megawatts.

As this is being written, publicity about nuclear power has centered on the politics of nuclear weapons made from reprocessed fuel. Iran might be

TABLE 6-4.   Sources of renewable energy

Source	% of total
Solar	1
Biomass	50
Geothermal	5
Hydroelectric	41
Wind	3

Source: U.S. Energy Information Administration 2007.

TABLE 6-5. Sources of electricity

Source	Megawatts per hour/10^6	%
Biomass	61.9	16.9
Geothermal	14.7	4.0
Conventional hydroelectric	270.3	74.0
Solar	0.6	0.2
Wind	17.8	4.9
Total	365.2	100.0

Source: U.S. Energy Information Administration 2007.

preparing a weapon, and North Korea may have tested one. In addition to these political concerns, there are long-standing concerns about safety, cost, and environmental issues.

In the 1960s and 1970s, it was thought that nuclear power would provide a continuing supply to supplant traditional sources, which were expensive and polluting. In 1984, *Engineering News-Record's* (*ENR* 1984) "man of the year" was lauded for the construction of Florida Power & Light's St. Lucie Unit 2 plant. This 802-megawatt plant was guided to construction in a record six years at a cost of $1.4 billion.

However, safety problems appeared on the radar screen with failures at Three Mile Island, Pennsylvania (1979), and Chernobyl, Ukraine (1986). The construction of new nuclear plants in the United States slowed to a halt after these failures. People became greatly concerned about margins of safety and the adequacy of the staff training and decisionmaking for and the monitoring and operational management of nuclear plants, as compared with conventional power.

Another issue has been the high capital cost of nuclear plants. The 1980s default of the Washington Public Power Supply System, caused by problems with its nuclear power plant construction, was the largest default of public bonds in the nation's history.

Nuclear waste has been another big issue, with a continuing national debate over an ultimate storage site for it. During the 1970s, it was thought that a study for a Waste Isolation Pilot Plant in New Mexico could be followed by the development of a permanent site at Yucca Mountain in Nevada. Now, some 30 years later, these plans seem to have ground to a halt, and most nuclear waste remains stored locally at the sites where it is generated. In addition to concerns about storage effects, transportation hazards loom large.

Energy Issues

In spite of Americans' best intentions, impacts on the environment from energy production remain a major concern. Today, global warming is the highest-profile issue, but past problems with air quality, water pollution, and land despoliation remain as well. Catalytic converters requiring unleaded automobile fuel began to be installed in 1975. The Clean Air Act amendments of 1990 entirely banned lead use, effective in 1996.

Our national dependence on imported oil supplies, particularly from unstable regions of the world, remains of great concern. From 1973 to 2006, U.S. dependence on imported supplies increased from 35% to 59% (Fialka 2006). Most of this dependence has resulted from our use of oil in the transportation sector.

The Corporate Average Fuel Economy (CAFE) standards established by the Energy Policy and Conservation Act of 1975 were supposed to put the United States on the road to energy efficiency. The CAFE standards set limits on miles per gallon of gas for new cars and initially reduced demand, but demand has now been increased by vehicles that are less fuel efficient. As of 2007, there was again interest in increasing and reforming the CAFE standards. If they are tightened, it will be the first change in more than 30 years.

Alternative fuels and new energy technologies are often in the news. Though they offer promising choices, their near-term impact is likely to be small. Alternative sources include general biofuels, methanol and ethanol, varied sources of natural gas, hydrogen, and other renewables.

Agriculture, Forest Products, and Aquaculture

Agriculture, forestry, and aquaculture are important users of natural resources. Of the nation's some 2.3 billion acres, forest use land accounts for 28.8%, grassland pasture and range land for 25.9%, and cropland for 19.5%. Thus, about 75% of all land falls into these three categories, which often involve an intensive use of land and water and have significant impacts on the environment and habitat. Decisions about them impact prime farmland and urban development, and their environmental effects can be pervasive. Though their economic heft is not a large percentage of overall gross domestic product (GDP), environmental managers often face decisions related to them.

The agriculture, forestry, and aquaculture sector accounts for a little over 1% of GDP. In its categorization according to the North American Industrial Classification System, it is divided into two parts: farm crop and animal production; and forestry, fishing, and related activities (U.S. Bureau of Economic Affairs 2007). Farm employment is tracked in a separate category than other industries and is difficult to summarize. As of May 2006, data from the U.S. Bureau of Labor Statistics (2007) indicated that there were 63,100 jobs in forestry and logging and 316,580 in support activities for agriculture and forestry. If all workers in basic agriculture and the processing of agricultural and forest products were to be added, these numbers would be much larger.

Agriculture

The importance of agriculture goes beyond its direct economic impact and often affects development and environmental decisions. By 1880, when the Bureau of the Census started keeping the statistic, the U.S. farm population was already under 50% of the total population, and farmers were 49% of the labor force. By 1990, the farm population was only about 1.1% of the total population, and farmers were only 2.6% of the labor force. The family farm has practically disappeared, and farm operators are competing more on a global stage with less protection than ever (Agriculture in the Classroom 2007).

The government's commitment to support agriculture is a long-standing policy issue. Farm bills are of the same order of magnitude as the national investments in transportation. These are expressions of the government's industrial policy toward agriculture. They include measures for price supports and farm subsidies, public health, security, agricultural research, energy, and food and nutrition, such as meals for poor schoolchildren (Rogers 2007).

Agriculture affects infrastructure and the environment in many ways. Examples are the transportation system to move commodities, nonpoint source runoff and water pollution, management of watersheds for water supply, farm-to-market road networks, the production of energy crops, and agriculture's role in preserving habitat for wildlife.

The Forest Products Industry

Forest products are used widely in the construction industry and for the production of pulp and paper. The residential construction market is a main driver of demand for lumber from the timber industry. The use of pulpwood by the paper industry is another driver, and forest products are also used for landscaping.

Watershed management is closely connected with forest practices. Forests create natural areas and habitats for wildlife. Logging occurs in watersheds and will increase sedimentation problems unless it is managed well. Trees absorb carbon dioxide and produce oxygen. Streams are sometimes used to float logs. When forests burn, it creates dramatic effects on the environment. Not only do wildfires destroy vast areas of forest, but the barren landscape is also vulnerable to floods, sedimentation, and mudslides. When you add up all environmental impacts of the forest products industry, they are huge.

Aquaculture

Aquaculture is another production industry that is closely connected with natural resources and the environment. It is a confined feeding operation, like that used to produce beef, chicken, or pork, except that it takes place in the water. Aquaculture is growing in importance and has important environmental effects because of the waste produced by the fish. If you add commercial and sport fishing to aquaculture, the overall fishing industry is much larger, but there are distinct differences between these and aquaculture.

Environmental Economics

Basic Issues

The differences between the fields of environmental economics and natural resources economics are in the use of terms and in the varying emphasis of each area. Natural resources economics deals more with the use of resources, whereas environmental economics deals more with the protection of the environment.

Environmental economics focuses on the economic tools to manage land, water, air, and habitat. It is concerned with controlling and allocating the use of resources through charges, incentives, and environmental laws and regulations. The specific issues with which it deals include land use control, water use allocation, water quality control, air quality control, and general environmental regulation.

The fundamental question in environmental economics is how to allocate scarce public resources. This question seems rational, but some people hold so strongly to environmental values that they do not like the idea of allocating the resources. Instead, they prefer to look for ways to preserve them. Regardless of value judgments, humans require environmental resources to survive, and the central question is how to use them to support all life and to sustain them for future generations.

"Sustainable development" is a useful concept to explain the need for balance in the use of environmental resources. Basically, it means using resources today in ways that will enable them to be preserved for tomorrow's generations. Definitions of sustainable development abound. The definition that most accept is by the Brundtland Commission (1987): "Development that meets the needs of the present without compromising the ability of future generations to meet their own needs." A few other definitions that were developed by candidates for a prize for achievement in sustainable development illustrate the many possibilities for defining the concept (Center for Global Studies 1993):

- Sustainable development is "a form of smart growth that employs the high-tech revolution and economic restructuring to manage all this growth in a more sophisticated manner that is ecologically benign."
- Sustainable development will "pass on to each generation a population level, a set of technologies, and a stock of fertile land and fossil fuels which would enable them to do at least what we have done."
- "The basic principle [of sustainable development] is that we live from flows, not from the stocks."

Environmental Indicators

The President's Council on Sustainable Development (1994) identified economic, environmental, and social indicators to measure sustainable development, as reported by the U.S. Interagency Working Group on Sustainable Development Indicators (1998). The group adopted a "pressure-state-response" framework to explain the indicators and an additional long-term accounting framework for it. "Pressures" are human activities that affect the environment, "states" are states of the environment or natural resources, and "responses" are societal responses to environmental concerns. This model emphasizes environmental issues more than economic and social issues.

The group's 40 indicators offer us a good framework for a "triple bottom line" assessment of infrastructure and environmental management programs (see Chapter 16). They were organized in 20 issue areas in three categories (Table 6-6). The economic indicators include the usual economic variables of efficiency and equity but add parameters such as employment, housing, and consumption. Environmental indicators include those you would normally expect, such as air and water quality, but add variables on the use of natural resources, ecosystem integrity, and the consumption of environmental capacity. Like environmental indicators, social measures are difficult to develop because there are so many possibilities.

Environmental Economic Analysis

For some types of environmental economic analysis, quantitative tools such as the benefit-cost ratio can be used. But because environmental issues include many intangibles, qualitative techniques, such as impact assessment, are also needed. For these qualitative tools, the basic task is to inventory resources such as plants, animals, water quality, air quality, and visual amenities and to assess how a proposed action will affect them.

TABLE 6-6. Indicators to measure sustainable development

Issue area	Indicators
Economics	
Economic prosperity	Capital assets, Labor productivity, Domestic Product
Fiscal responsibility	Inflation, ratio of federal debt to gross domestic product (GDP)
Science and technology	Investment in research and development as a % of GDP
Employment	Unemployment
Equity	Income distribution, poverty levels
Housing	Homeownership rates, percentage in problem housing
Consumption	Energy and material use per capita and per $ of GDP
Environment	
Status of natural resources	Cropland, soil, water use, fisheries, timber use
Air and water quality	Surface water quality, air quality nonattainment
Contamination	Contaminants in biota, superfund sites, nuclear waste
Ecosystem integrity	Terrestrial ecosystems, invasive alien species
Global climate change	Greenhouse gas emissions
Stratospheric ozone depletion	Status of stratospheric ozone
Social	
Population	U.S. population
Family structure	Children in one-parent homes, births to single mothers
Arts and recreation	Outdoor recreation, participation in the arts
Community involvement	Contributing time and money to charities
Education	Teacher training, educational achievement rates
Public safety	Crime rates
Human health	Life expectancy at birth

Source: U.S. Interagency Working Group on Sustainable Development Indicators 1998.

Environmental impact analysis and environmental impact statements constitute a way to compare projects. The 1970 National Environmental Policy Act ushered in the long-standing requirement for them.

Social impact analysis is another way to compare projects. It is not used as much as economic and environmental analysis tools, and its techniques are not as standardized.

Environmental Management

The debate over whether it is better to regulate the use of resources or let the market allocate them continues. In developing the Clean Water Act, for example, proponents of pricing said that the environmental capacity to assimilate wastes should be set and the right to pollute should be sold. Proponents of regulation said that this would not work and that the government should adopt a command-and-control approach.

The economic approach to allocating pollution rights through the market has strong intellectual underpinnings. However, the practical difficulties of administering such a system are formidable. Neither the economic nor the regulatory approach works perfectly. Though the regulatory approach is most common for environmental cases, allocation by pricing can be used for some cases, such as the sale of air pollution rights. Research is proceeding on how to allocate other environmental resources, such as nutrients through trading schemes.

Environmental economics intersects and merges with politics and ethics in many ways. The tension between private market and government solutions to environmental problems requires other types of solutions than pure command-and-control regulatory approaches.

Water Resource Economics

Main Issues for Water Decisions

Among environmental questions, water issues garner perhaps the most attention from infrastructure and environmental managers. Water issues spill over into many other environmental decisions and vice versa. Water is scarce in some regions and will cost more in the future, as new supplies diminish and conflicts increase. This subsection provides an overview of water economics, including supply and demand, how water is used and valued, and how the water industry operates.

Water management involves natural resources, government, regulation, and private sector activity. For the most part, water economics involves public sector issues, because the main questions involve costs and benefits that lie outside pure business considerations.

The fact that water is, on the one hand, a commodity to use and, on the other, an environmental resource to be protected creates a dilemma and requires a balance among uses and protection. You will encounter this dilemma in many places, not least in the false perception that engineers always want to exploit the water and environmentalists always want to protect it. As with many complex issues, the truth is usually in between.

Another dilemma is the perception that water should be free. Though nature provides water as a "free good," the infrastructure to manage it is costly, and thus its commodity value and cost are significant. Also, systems for administering water and for protecting the natural water environment are costly.

This dilemma leads to the problem that if something costs little, people do not value it and will waste it. Tracy Mehan (2007), former assistant administrator for water of the U.S. Environmental Protection Agency, told a story of how a nun protested a rate increase by saying that since God provided the water, it ought to be free. The water manager replied, "Sister, we agree the water should be free, but who will pay for the pipes and pumps?" In other words, the cost of water is due to infrastructure, not the commodity.

People are often confused about whether water is a "public" or "private" good because, looking at the extremes, they think that you can leave all water services and activity to private companies, or that water ought to be free. In this debate, one side wants water for the public trust and for free distribution, while the other sees it as an economic commodity. Thus, water has attributes of both a public good and a private good. Some water services can be handled by the private sector, but water is also a public good with respect to functions such as environmental water quality management. Water cannot be free because it must be collected, processed, and managed. These services are amenable to provision by the private sector or by public sector utilities.

The Financial Cost of Water Versus Its Opportunity Cost

One way to explain the false perception that water should be free and to clarify the difficulty in valuing it is to look at the financial cost of water versus the opportunity cost of water in use. The financial cost of water comprises the cost of the pipes and pumps and other infrastructure, whereas the opportunity cost of the resource is its value in its highest-value alternative use. For example, if you divert water for irrigation, the financial cost might be, say, $1,000 an acre-foot. This might seem attractive for investment, but if the water could be used later for urban use and have a value there of $10,000 per acre-foot (based on willingness to pay), then the allocation decision looks much different. According to John Briscoe (2008) of the World Bank, the correct valuing of the opportunity cost is the key to establishing a viable water market.

National Water Industries

National water policies depend on the scale of the country. U.S. water policy is more complex, for example, than water policy in a small country. National water policy is only meaningful if it leads to action. In the United States, national policy under the Clean Water Act had big effects, but the Water Resources Planning Act of 1965 ended in inaction.

The size and scale of the country determine the political distance from policy level to local implementation. In a highly centralized small country, water use decisions might be made in the capital, whereas in a giant country like the United States or China, they will be more regionalized. Regional and local water use decisions can be very significant—for example, those for the water supply of Southern California.

The degree of privatization or state control of water is another important variable. This requires decisions about public or private ownership of water resources and about the regulatory structure of the water industry.

Water industries vary as well, depending on the extent of a nation's industrialization and urbanization and of its reliance on irrigated agriculture. For example, water management in Egypt is tightly interwoven with food production, but in Finland irrigation is not an important factor.

The Microeconomics of Water

The microeconomics of water, focused at the utility level, address issues such as allocation, pricing, incentives, equity, and policy, including

- supply, or how a utility decides on investments in water system capacity;
- demand, or the mixture of pricing and controls that will maximize the efficiency of water use;
- incentives for conservation and water quality management;
- equity in the allocation of water and water services and subsidization; and
- economies of scale, or the optimum organization of water services.

Finance in the Water Industry

Financial decisions in the water industry deal with questions such as privatization, water pricing, cost pressures, accounting and control, the need for subsidies, government grants, revenues, and access to capital. The water industry has widespread impacts on the economy, society, and environment. Its statistics are embedded in public and private reports of government, construction, utilities, and services.

The water supply and wastewater sectors have revenues of about $40 billion each. Statistics are not maintained for stormwater and flood control activities, which are largely in the government sector. Irrigated acreage and agricultural statistics are maintained by the U.S. Department of Agriculture and state governments. However, expenditures for irrigation and drainage services and equipment are not kept by any government agency. The economic impacts of water quality management, fish and wildlife, hydroelectric power, navigation, and recreation are large but difficult to measure and report. Decisions by regulators are important to the economy—for example, imposing stricter wastewater treatment rules. Also important, but difficult to measure, is the work of water industry support organizations, including equipment suppliers, contractors, consultants, lawyers, associations, interest groups, and educators and publishers.

Flows of Funds in the Water Industry

The water industry includes sectors for water supply, wastewater management, irrigation and drainage, environmental water, and others. In the United States, it accounts for about 1% of GDP and it requires about one million employees, including employees of suppliers to the water providers (Grigg 2007).

In the water industry, funds flow from taxpayers and ratepayers through utilities, governments, and businesses, and are recycled to suppliers, employees, and others who provide the goods and services needed to keep the water flowing. Figure 6-6 shows how these flows originate from households and businesses, pass through the sectors, and result in water-related services needed by the economy and society. The payments can include charges for services and taxes to cover broad public interest issues.

Western Water Rights

In the western states, water has a commodity cost because the right to use water is a property right that can be bought and sold. The sale price for a 1,000 acre-feet water right might be, for example, $2,000 per acre-foot, for a total value of $2 million. This might be enough water for a town of 4,000 people or a farm of 500 acres, more or less. The $2 million in the bank at 5% would earn interest of $100,000 per year, but it would normally bring less if it was leased out or rented. This paradox occurs because the commodity value of water is based on average yields and the lease value varies with shortages and surpluses. For these values to be used to make deals, a working market must be available.

Incentives and Equity Considerations

Conservation and water quality management require economic tools to promote them. Positive incentives include pricing and rewards, whereas

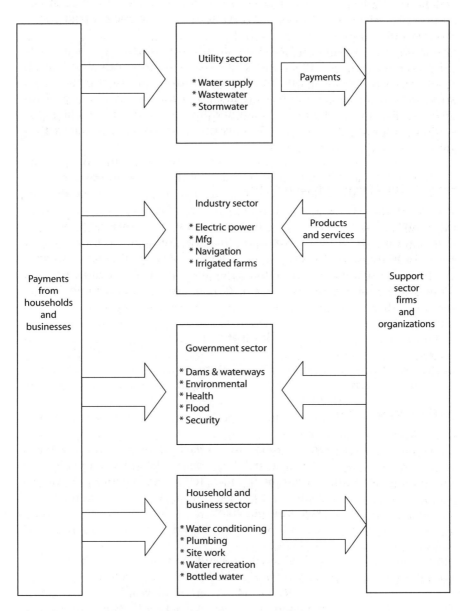

FIGURE 6-6.   The flow of funds in the water industry

negative incentives involve penalties and cost surcharges. Demand management involves the set of tools and methods that is used to forecast and regulate demand for water and water services—mainly pricing, regulations, and incentives.

Economics explains equity, or the distribution of costs and benefits, to different sectors of society. For example, wealthy citizens can afford all the water they might require, but lower-income citizens might struggle to pay minimum utility bills. Where is the fair point for the distribution of the water supply and charging for it? Should lower-income citizens be subsidized? If so, who should pay? These issues are considered in rate-setting policies.

Economies of scale occur when, as production rises, the cost per unit of production falls. From an economic standpoint, economies of scale make sense, but there may be cases in water management where they are not appropriate. For example, if water is priced at the "cost of service," one could argue that as more is used, it ought to cost less. However, to promote conservation, the opposite would be the case. Another economy of scale might be to build large, centralized treatment plants, but that might prevent increasing redundancy by distributing treatment plants throughout systems. Thus, economies of scale are only one of many issues in decisionmaking.

References

Agriculture in the Classroom. (2007). A history of American agriculture. http://agclassroom.org/textversion/gan/timeline/temp_decade.htm. Accessed June 30, 2009.

Briscoe, J. (2008). "Valuing water properly is a key to wise development." *Wall Street Journal*, June 23.

Brundtland Commission. (1987). *Our common future: Report of the World Commission on Environment and Development.* World Commission on Environment and Development. Annex to General Assembly Document A/42/427. United Nations, New York. http://www.un-documents.net/wced-ocf.htm. Accessed June 30, 2009.

Center for Global Studies. (1993). "Towards understanding sustainability." *Woodlands Forum*, 10(1).

Denver Post. (2007). "Coal mine safety is still inadequate." August 21.

Economist. (2007). *Pocket World in Figures.* Economist, London.

Engineering News-Record (ENR). (1984). "Man of the year." February 9.

Fialka, J. J. (2006). "Energy independence: A dry hole?" *Wall Street Journal*, July 5.

Grigg, N. (2007). "Water sector structure, size and demographics." *Journal of Water Resources Planning and Management* (ASCE), 133(1), 60–66.

Lubowski, R., Vesterby, M., Bucholtz, S., Baez, A., and Roberts, M. (2007). *Major uses of land in the United States, 2002.* Economic Research Service, U.S. Department of Agriculture. http://www.ers.usda.gov/publications/ eib14/. Accessed July 5, 2007.

Mehan, T. (2007). "God gave us the water, but who pays for the pipes?" *Water & Wastes Digest,* 47(5), 1.

President's Council on Sustainable Development. (1994). Brochure for workshop on challenges to natural resource management and protection of the Colorado River Basin, University of Nevada, Las Vegas. December 12.

Rogers, D. (2007). "House passes big farm bill." *Wall Street Journal,* July 28–29.

Smith, S. (2007). Minerals information: Statistical summary. http://minerals.usgs.gov/minerals/pubs/commodity/statistical_summary/. Accessed July 2, 2007.

U.S. Bureau of Economic Affairs. (2007). North American Industrial Classification System. http://www.bea.gov. Accessed July 15, 2007.

U.S. Bureau of Labor Statistics. (2007). Current employment statistics. http://www.bls.gov/ces/#data. Accessed August 20, 2007.

U.S. Energy Information Administration. (2007). Energy basics 101. http:// www.eia.doe.gov/basics/energybasics101.html. Accessed February 15, 2007.

———. (2008a). Basic energy statistics. http://www.eia.doe.gov/basics/quick oil.html. Accessed May 16, 2008.

———. (2008b). Coal. http://www.eia.doe.gov/fuelcoal.html. Accessed May 17, 2008.

U.S. Environmental Protection Agency. (2007). Frequent questions about mercury. http://www.epa.gov/mercury/faq.htm. Accessed July 15, 2007.

U.S. Interagency Working Group on Sustainable Development Indicators. (1998). *Report of the Interagency Working Group on Sustainable Development Indicators.* U.S. Government Printing Office, Washington, DC.

U.S. National Technical Information Service. (2007). Bureau of Mines publications. http://www.ntis.gov/products/specialty/bom.asp?loc=4-5-1. Accessed February 17, 2007.

7

Utility Economics

Public Utilities

Around 1900 it became clear that new services for electricity and the telephone involved management arrangements that were different from those in a purely competitive marketplace. These services required large investments and had the characteristics of natural monopolies. In their early days, they experienced continuing technological change because many discoveries were being made. Later, water and wastewater came to be known as utilities as well, and the concept of the "public utility" took shape.

A utility is defined as a service provided to the public, such as electricity, natural gas, water supply, or solid waste collection. These services can be financed through user charges rather than tax subsidies. As government seeks to be more efficient, there is also a trend toward defining other public works services as utilities, such as stormwater and street maintenance.

Today, many decisions by infrastructure and environmental managers relate to energy, water, and waste management utility work. Public utilities that provide these services employ more than a million workers in the United States, require large inputs of capital and operating funds, and make decisions with significant economic, social, and environmental consequences. This chapter presents the economic principles that form the basis for utility management.

Many economic principles of utility management have been developed, and many have been tested in the legal arena. Initially, these came mostly from the electric power and telecommunications sectors, but now each utility sector has its own history and regulatory arrangements. The starting point for understanding the economics of public utilities is to consider a classification scheme for them.

A Classification Scheme for Public Utilities and Public Services

A classification system is used to distinguish public utilities from other private or government services. In government, what makes electric power service a utility, whereas the Internal Revenue Service is not? On the private side, if electric power is a utility, why would a gasoline station not be one?

Figure 7-1 provides a basic answer to these questions by presenting three aspects of services: a product or service of the private sector, a public utility, and a pure government service. Utility economics explains why these three categories should be differentiated. The basic reasoning is that the market is free to provide almost any good or service demanded by the public. That is what the term "free enterprise" means. Some basic services may involve monopoly franchises and cannot be offered easily by competitors in the market. These might become utilities. Government services are required in certain critical areas, such as health and safety, and do not involve either market goods or utilities.

Although the three categories are clearly different, it can be hard to draw the lines between them. Clearly, the market can handle purely discretionary services, and only the government should have an army or collect taxes, but what about services that fall between these extremes? How do we decide if a service qualifies as a utility? And even if it does qualify, how do we decide whether the government should manage it or not?

Table 7-1 shows a scheme for discriminating among services on the basis of whether they are essential public services, whether they can be measured and rationed or not, whether they have substantial public benefits or not, whether they can be offered by private firms or only by the government, and whether the service is diminished by use or not. Yet even this scheme, with its several discriminating variables, is not always able to draw sharp distinctions. For example, electric power is an essential public service and can be measured, so it is in the public utility category. However, experience shows that it can be offered by either the government or private firms.

The classification scheme outlined in Table 7-1 presents many judgment-based choices, such as "Is a service essential and not discretionary?"

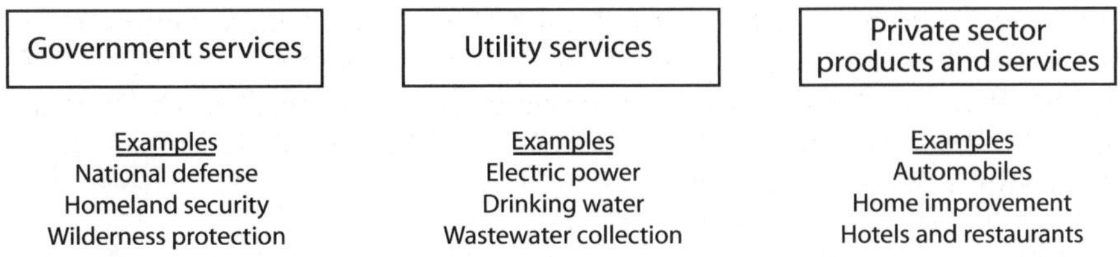

Government services	Utility services	Private sector products and services
Examples	Examples	Examples
National defense	Electric power	Automobiles
Homeland security	Drinking water	Home improvement
Wilderness protection	Wastewater collection	Hotels and restaurants

FIGURE 7-1. Government services, public utilities, and private sector products

TABLE 7-1. Classification scheme to identify public utilities and public services

Category	Examples
Public utility–type goods: offer essential public services that can be measured and rationed by charging schemes	Airports, water supply, electric power, gas supply, transit services, and water pipelines.
Private goods with important public purposes: provide benefits to society as a whole but can be offered by private firms	Sewerage services, trash collection and disposal, industrial waste disposal, and toll roads and bridges.
Services where public purposes dominate: one person's use of the service does not diminish its availability to others	Water effluent control and water quality management, air pollution control, and highway usage control

Source: Examples drawn from Mushkin 1972.

Water supply is essential, but what about trash collection? Is the service a public good, and does it serve important public purposes? Does service to one person affect another's access to it, or is a person's use of it interdependent with use by others? Can it be unbundled, rationed, and measured? For example, the generation, transmission, and distribution of electric power can be unbundled. Is it therefore a "natural" monopoly, or can competition be open? And finally, does it require some kind of government regulation, and if so, why?

Not all the services with which infrastructure and environmental managers deal have been classified rigidly. As shown by the classification given in Table 7-2, some meet the definition of utilities more than others. Though the utility status of these services might seem clear, it is open for debate. For example, water supply is essential to life, and even though some people can live in rural areas and drill wells, most people depend on piped water. Water supply requires regulation for safety and other purposes. If one person uses more water, less is available for other purposes. So, strictly speaking, water is not a public good, although most people would consider it as one. Piped water supply is a natural monopoly, and it can be rationed and measured.

Wastewater service is also essential, but in different ways than water supply. Even though some people can live in rural areas and have septic tanks, most people discharge to public sewers. Wastewater requires regulation for environmental and other purposes. If one person discharges more effluent, it affects others by taking up the capacity of infrastructure systems, so it is

TABLE 7-2. Classification of services by three characteristics

Service	Essential	Require regulation	Monopoly
Water supply	Yes	Yes	Yes
Wastewater	Yes	Yes	Yes
Electric power	Yes	Yes	Yes
Gas	No	Yes	Yes
Transit	No	Yes	No
Solid waste	Yes	Yes	No
Storm drainage	No	Yes	Yes

not really a public good. Wastewater collection is a natural monopoly and can be measured. It can be rationed, but only through the water supply.

In today's world, electric power is essential. Although some people might get "off the grid," most people depend on an electric power utility. Electricity requires regulation for rates and other public purposes. If one person uses more electric power, it will affect others' uses, although not by much. Electric power distribution is a natural monopoly and can be measured, but generation and transmission can be unbundled from it.

If all-electric living is possible, natural gas service is less essential. Gas requires regulation for rates and safety. If one person uses more gas, it affects the supply for others. Gas distribution is a natural monopoly and can be measured, but generation and transmission can be unbundled from it.

Transit is not essential to many people because alternative modes are available. It has social purposes, however, because the alternatives are not available to everyone. Transit requires regulation for rates and access. If one person uses more transit, it has little effect on others' uses, other than crowding. Transit is not a natural monopoly, but if it is not regulated, then service may diminish due to unprofitability. It can be measured and charged for.

Theoretically, solid waste collection is not essential because a person could reduce their waste to zero. In a practical sense, however, it is essential. Solid waste collection requires regulation for service levels, rates, and other purposes. If one person uses more collection capacity, it has little effect on others' uses. Solid waste is not a natural monopoly. It can be measured and made a commodity.

Storm drainage is similar to wastewater service, but it is not as essential. Streets carry away much of the storm drainage, and additional systems are not always needed. They provide "convenience" type services more than wastewater does. Stormwater requires some regulation, but less than wastewater. If one person's property discharges more drainage, it affects others by using the capacity of the common drains. Stormwater is a natural monopoly and can be measured. This is the basis for the creation of "stormwater utilities."

In the past, telecommunications and cable television were considered utilities. Now, this designation seems to be called into question due to the increase in competition among them that was ushered in by new technologies and deregulation. Telephone service is essential for modern life, but there are now different forms of it. It requires regulation to allocate the frequency spectrum and for some other issues. If one person uses more of it, it has little effect on others, but it cannot be considered a public good. Phone service is no longer a natural monopoly. It can be measured and unbundled.

Cable television and its competitors are not essential because many people live without them. They require regulation to allocate frequencies and local service areas, but increasing competition among modes is changing the need for regulation. If one person uses more of this service, it has little effect on others, but it cannot be considered a public good. It is not a natural monopoly. It can be measured and unbundled.

After looking at these individual cases, it is apparent that to be a public utility, as opposed to a discretionary market service, the service should be essential, provide at least some important public purposes, and deliver a commodity that can be rationed and is not changing constantly through innovation and entrepreneurship. Water supply, electricity, and wastewater services clearly fit these criteria.

Infrastructure services that can be financed through user fees can logically be called utilities. Some, like flood control, are usually financed through taxes and are less like utilities and more like government services. However, stormwater has been shown to be amenable to utility financing. However, user fees for services like stormwater are harder for the public to understand than are fees for commodity-type services, but with pressures to limit taxes, the utility approach to financing these services may be the only option.

Utility Regulation

Utility regulation has the general goal of protecting the public interest, and this takes on different facets for particular utilities. To illustrate, Table 7-3 provides examples of the public purposes for regulating two types of utilities, the water supply and electric power. Within these regulatory categories, you can find many other examples. For example, if an electric power utility could cherry-pick its customers and was only motivated by profit, it would serve some but not all the customers and access would be a problem. So, in exchange for the right to serve customers, utilities might be required to serve

TABLE 7-3. Examples of the public purposes for regulating the water supply and electric power

Public purpose	Water supply	Electric power
Health and safety	Safety of drinking water; maintain water quality in streams	Electric safety codes and construction rules
Environment	Instream flows for fish and wildlife protection	Air quality rules for coal power plants
Resource access	Enforce legal water rights	Enforce permit conditions
Finance	Control rates of private water companies	Control rates of private electric companies
Service access and quality	Guarantee access to water; maintain adequate water pressure	Guarantee access to electric power; maintain electric power reliability

all customers and to build out to the limits of a service area. Other issues include the cost of service and differential costs.

The Utility Industries

When Wall Street refers to "utilities," it is mostly talking about electric power. The Dow Jones Utility Index is based on 15 utilities, all of which are electric power businesses except one, a natural gas utility. Of these 15, all are "large cap," except two "mid cap" businesses (Dow Jones 2007). Natural gas as a utility industry also has visibility with investors, but water utilities have a lower profile.

Electric Power Utilities

According to the Edison Electric Institute (2006), electric power is about a $300 billion per year industry and employs about 400,000 workers.[1] Clearly, this utility industry has massive impacts on the economy, as well as providing essential energy inputs to homes, businesses, and government. It is larger than the pharmaceutical or airline industries.

---

[1]The Edison Electric Institute data are drawn mainly from its 2005 financial report.

The electric power industry started in the 1880s, when Thomas Edison saw business opportunities in lighting streets with his new invention, the light bulb. He built central power stations to provide direct current power, and his Edison Electric Light Company eventually led to the formation of the General Electric Company. The new electric power industry attracted many players. One of them, Samuel Insull, immigrated from England to be Edison's secretary. Insull became a utility baron in 1920s Chicago, and he thought that by agreeing to regulation, he could avoid being taken over by public power interests. The regulation by public utility commissions dates to this era (Wasik 2006).

The work of Edison and Insull marked the start-up period for electric power. As the industry grew, it was shaped by business and government initiatives. Along the way, utility barons created large businesses and the government organized the Tennessee Valley Authority. After that, the industry expanded through rural electrification, the growth and decline of nuclear power, and a continuing increase of demand. During the 1990s, the country flirted with electricity deregulation, with some unbundling of generation from transmission and distribution.

Today, four basic types of organizations provide electric power: investor-owned utilities; public systems owned by the federal government; public systems owned by states, municipalities, or utility districts; and cooperatives. The largest investor-owned utilities are familiar names: Pacific Gas and Electric, Commonwealth Edison, the Southern Company, and American Electric Power. Federal systems revolve around six agencies that market power: the Bonneville, Alaska Southwestern, and Southeastern Power administrations; the Bureau of Reclamation; and the Tennessee Valley Authority.

In the United States, electric power is mostly provided by publicly regulated private companies that are owned by shareholders (72.2%). The rest of the nation's electric power is provided by municipal and political subdivisions (13.6%) and by cooperatives (12.3%) (U.S. Energy Information Administration 2007). For example, in my city, electric power is purchased and distributed by the Fort Collins Electric Utility, but neighboring cities get power from Xcel Energy and rural areas get it from cooperatives. Fort Collins entered the power business in 1935, when it established its light and power department. The power superintendent used the slogan "Electricity is cheap: use it freely" (City of Fort Collins 1985).

The U.S. Rural Electrification Administration has helped to electrify rural America since the 1930s and still provides low-interest loans. A typical rural cooperative is dependent on subsidized loans and low-cost federally provided power.

The larger electric power companies are the 203 shareholder-owned utilities. Municipal government electric systems are the largest number of utilities, at 1,874. However, 1,688 nonutility generators produced 35% of all power in 2005. These include cogenerators and small and/or independent producers. There were also another 1,433 energy service providers, mostly small businesses. Some 870 cooperatives produced power. And there were 133 public power districts and another 40 state and federal producers (Tennessee Valley Authority and federal power administrations).

Electric utilities account for almost 40% of all primary energy demand, up from about one-third in 1982 (U.S. Energy Information Administration 2007). Most customers are residential (87.3%), but the 0.5% of customers who are industrial use more than one-third of the energy produced. Commercial users account for 12.2% of customers, and municipal transit systems represent less than 0.1%.

Electric power systems place a substantial call on capital resources to expand and maintain their production, transmission, and distribution facilities. They account for a large share of new industrial construction, corporate financing, and common stock issuance among industrial companies. Capital spending was $46.5 billion in 2005 and was projected to reach $60 billion in 2006. This is smaller than the transportation sector but much larger than the water sector.

Regulation in the electric power industry is split between the federal role, centered in the Federal Energy Regulatory Commission, and the role of the states, whose public utility commissions oversee investor-owned utilities.

Natural Gas Utilities

The production, transmission, and distribution of natural gas also constitute an important energy utility service. Gas supplies about 22% of primary energy in the United States.[2]

The natural gas industry evolved along with the nation's energy needs. For example, Baltimore Gas & Electric (2007) claims to be the nation's first gas utility, with street lighting service dating back to 1816. In 1916, its natural gas and electric services were merged, and after World War II, it took its current name. The Natural Gas Act of 1938 initiated federal regulation of the industry to oversee interstate pipeline companies and to give the Federal Power Commission (now the Federal Energy Regulatory Commission) the authority to set rates for interstate transmission and sales. The act does not regulate local gas distribution utilities (U.S. Energy Information Administration 2007).

The natural gas industry is not very integrated, and thus it has separate exploration and production companies, transportation pipeline companies, and local distribution utilities. Some 8,000 producers provide natural gas in the United States, ranging from very small operators to large integrated

---

[2]Most of the information in this section is from the Natural Gas Supply Association (2007).

global companies. There are more than 580 natural gas processing plants that in 2000 processed about 17 trillion cubic feet of gas and extracted more than 720 million barrels of natural gas liquids. About 160 transmission pipeline companies operate in the United States, with more than 285,000 miles of pipe. This includes 180,000 miles of interstate pipelines with a capacity of about 119 billion cubic feet of gas per day. Also, 114 storage operators in the United States have 415 underground storage facilities. Some 1,200 local distribution companies operate in the United States, with about 833,000 miles of distribution pipe. These mostly have monopoly status, but some states require options in distribution.

The largest natural gas production companies, such as Exxon, Texaco, and Mobil, also produce oil. Large transmission companies include El Paso Natural Gas Company, Columbia Gas Transmission Corporation, Tennessee Gas Transmission Company, and the Natural Gas Pipeline Company. Large distribution companies include Southern California Gas Company, Pacific Gas and Electric Company, InterNorth Incorporated, and Consolidated Gas Supply Corporation.

Technology in the natural gas industry focuses on the movement of gas. In 1930 pipelines were limited to about 20 inches, with pressures up to 500 pounds per square inch, but the sizes and pressures are much greater now.

In the past, the wellhead and wholesale prices of natural gas were federally regulated, and its local distribution was regulated by state public utility commissions. Now, more ownership pathways exist for gas to move from producer to end user. Wellhead prices are no longer regulated, but transmission remains regulated by the federal government. Natural gas marketers facilitate its movement from the producer to the end user. In 2000, about 80% of natural gas in North America was handled by them.

The issues faced by the natural gas industry include energy supply, rate regulation, and the condition of the transmission and distribution infrastructure.

Water, Wastewater, and Stormwater Utilities

The water supply, wastewater, and stormwater utilities are part of the larger water sector, which also includes agricultural water, flood control agencies, dam owners, and environmental water uses (Grigg 2006). Water supply utilities evolved over a longer period than electric and telephone utilities because its technologies were not based on sudden technological discoveries. Early people learned to divert water and bring it to settlements and farm fields through canals, aqueducts, and crude pipes. By the seventeenth century, cast iron pipes were delivering water to many cities around the world.

Organized water utilities emerged as U.S. cities developed. In these early days of providing water supply in America, every city had a unique story

(Armstrong 1976). Philadelphia initiated its water supply system in 1798 after a yellow fever epidemic. It used public and private pumping facilities powered by horses. Other large cities, such as New York and Boston, followed Philadelphia's lead, and developing water systems became an important civic accomplishment. In parallel, the craft of plumbing evolved during the latter part of the nineteenth century.

Safeguarding public health became an important objective of drinking water systems. Water treatment systems were developed only after the field of microbiology emerged in the nineteenth century. In 1974 the Safe Drinking Water Act was passed to usher in stricter controls on the safety of drinking water.

Before about 1900, essentially all water services were private. During the twentieth century, a number of private water companies were assembled. Along the way, mistrust of private sector "trusts" and "robber barons" led to pressure for government involvement, and the trend shifted toward public sector management. After about 1980, the pendulum swung back. Internationally, under Margaret Thatcher's leadership as prime minister, the United Kingdom privatized its water industry. Today, there is much interest in privatization, although it is not universally favored.

Wastewater systems began with in-house water supply and water closet systems, which overloaded cesspools and privy vaults. Storm drains were necessary to keep cities from flooding, and it was easy to dump wastewater into them. Combined sewers emerged as a solution to carry away domestic waste, only to create their own problems. The discovery of the water seal for building drains made in-house plumbing socially acceptable, whereas without the seals the odor was unacceptable. Mixed sewers caused great water pollution. As cities developed in the twentieth century, the trend shifted toward separate sewers. After the Clean Water Act was passed in 1972, large investments were put into the construction of wastewater treatment plants. Until their improvement was spurred by that act, wastewater services did not involve much sophistication. Only since then has wastewater emerged as a utility—and only with government encouragement. Environmental enforcement and more stringent clean water programs followed. Today, wastewater is a recognized utility service in almost every city.

Urban storm drainage is often linked with flood control as a service, but in reality it involves different systems and goals. Because streets need drainage to maintain their viability, it was natural that with urbanization, ditches and drains would be created. During the urbanization phases of the twentieth century, storm drainage systems became more extensive, and by the 1980s stormwater utilities were formed, mainly as a way to find revenues other than general taxes to support their infrastructure. You can make the argument that stormwater is a utility service, rather than something provided for the general good. Charges for the service can be based on the quantity of stormwater generated by the impervious surfaces of land developments. At about the same time, the Clean Water Act ushered in the regulation of some stormwater discharges, and stormwater management became a more recognized service in urban areas.

Government has a large role as regulator and as support provider through water agencies. The support sector of the water industry, which is also large and complex, includes associations, professional service firms, suppliers, knowledge sector, advocacy groups, construction contractors, and financiers and insurers.

The water sector can be measured by revenues, number of establishments, and number of employees. My estimate of the water industry's annual revenue for water, wastewater, and stormwater is $100 billion, based mostly on association statistics and local government budgets (Grigg 2006).

It is estimated that there are some 60,000 community water systems and 50,000 wastewater management units in the nation. Some of these are so small that it is a stretch to call them utilities. Stormwater units are mostly found inside city government organizations.

The size of the nonutility part of the water industry is more difficult to identify and measure than are the utilities. A few of the categories it encompasses are nonutility water districts, including irrigation districts; industrial self-supply and wastewater management; federal, state, and regional water agency budgets; and maintenance fees for building plumbing systems. Adding up all these, my estimate of the combined size of the water industry is about 1% of the U.S. gross domestic product.

Solid Waste Management

Solid waste management is divided into collection, transfer, disposal, recycling, and hazardous waste disposal services. It is a necessary service with many health and environmental impacts.

Early solid waste management had an ignoble beginning, when even wealthy people tolerated garbage dumps near their homes. The earliest organized solid waste service in the United States may have been Benjamin Franklin's 1792 initiative to have his servants collect waste and dump it in the Delaware River. In 1844, the city of Washington began the private collection of waste from homes to supplement its ongoing street-cleaning services. In that era, cities were unhealthy places because modern sanitation and public health practices had not yet developed. With the large rise in urbanization between 1860 and 1910, and the emergence of public health as a professional field, solid waste collection and street cleaning became imperative. Today, solid waste management and street cleaning are considered essential public services (Armstrong 1976).

In its early form, people saw solid waste management as including only disposal, but it has evolved beyond that. Today, it handles many kinds of waste and has many more options available. Federal regulation of the field is mainly through the Resource Conservation and Recovery Act of 1975. Numerous other laws also govern the disposal of solid and hazardous wastes, for example, the Hazardous Materials Transportation Act and the Superfund administered by the Environmental Protection Agency.

Most solid waste is still disposed of in landfills using traditional technologies, although this is an improvement over earlier open dump and burning approaches. However, disposal agencies are running out of space, and the landfills contaminate groundwater unless protection is provided. Recycling and using waste in energy-generating plants offer some relief from the problems with landfills, but they need much more emphasis and development.

The collection of solid waste is labor intensive and accounts for some 80% of the total cost of solid waste management. Efforts have been made to automate collection to reduce labor requirements, but these may have raised efficiency about as much as they can. A collection enterprise might supply large containers for curbside pickup so even older citizens can place their own waste at the curb. These containers might be lifted into the truck with a mechanical arm so that even one person can operate a truck.

Solid waste management has become a major business sector. Privatization is common in collection services, but many local governments still operate their own services, usually in a public works department.

Transit Service

Transit services are sometimes classified as utilities. By the 1830s, transit service had emerged with the omnibus, which was pulled by horses. Later, horses pulled horsecars on tracks. Steam engine technology and cable cars offered additional alternatives, and then electric trolleys emerged after 1880. Later, subways, surface railways, and bus systems began serving urban areas, but ridership fell off with the rise in automobile travel. Today, mass transit is a small but important part of the national urban transportation system (Armstrong 1976).

Communications Services

Communications services have changed rapidly, and it is hard to call them utilities any more. When telephone service was new, service consisted of a telephone monopoly that provided a phone, a wire, a switchboard, and the ability to talk with other people. Local service was by party line until the capacity increased enough to give each person a private line. Long-distance service would be expensive, if it was available. The integrated telephone monopoly did research, built systems, sold equipment, and provided the full range of services at regulated prices.

The telephone industry became a public utility through development of the Bell System. After an initial skirmish with Western Union, the American Telephone and Telegraph Company (AT&T) had a virtual monopoly on telephone service through its regional companies. AT&T delivered local and long-distance phone service, manufactured equipment, and did research through Bell Labs to improve its systems. Only in the 1980s, with a federal court order, did this tight monopoly on phone service unravel.

Once the telephone monopoly was broken up, communications became much more competitive. Mobile telephones, wireless services of different kinds, phone calls over the Internet, and other provider channels are now available. Phone companies compete with cable services to provide bundled services of land line, cell phone, Internet, and broadcast television.

Cable television was initiated in the late 1940s in Pennsylvania, and it was originally known as community antenna television, or CATV (About Inventors 2007). Today, there are more than 100 cable networks in the United States, and new technologies and competitive business models continue to unfold. Whether cable TV will continue to be considered a utility is questionable, given the choices in delivery modes.

Communications services are still regulated, but the rules must change to adapt to the new technologies and business systems. The federal and state governments are involved in regulating aspects of the business that range from the rights to use the radio frequency spectrum to the service areas for communications companies. In the case of cable TV, it has been regulated by some 33,000 local franchises, about one for every city and town in the United States. In 2006 Congress debated a Communications Opportunity, Promotion, and Enforcement Act, which would introduce a national franchise system. Such a bill would remove barriers to competition, as the Federal Telecommunications Act of 1996 did for telephone and video service. This approach may be a long time in coming, because cities are not sure they want the Federal Communications Commission to regulate these local issues (*Denver Post* 2006).

Capacity Expansion in Public Utilities

Planning for capacity expansion is a common issue in public utilities, and it requires using economic decision tools such as cost-benefit and rate-of-return analyses. As the population increases, the demand for utility services usually increases as well. Take water supply for example. If per capita demand remains constant, then total demand for water will rise with the population increase. If the infrastructure facility required is a water treatment plant, then the required capacity constantly increases.

It makes sense to expand infrastructure in stages rather than continually, so the capacity expansion problem in economics has emerged to explain how to optimize overall outcomes by applying economic decision tools. The capacity expansion problem can be framed by the diagram given in Fig. 7-2, which illustrates constantly increasing demand being met by capacity increases that occur from time to time.

Deregulation

Utility monopolies create a drag on competition, productivity, and efficiency, even when they are regulated by utility commissions. Thus, deregulation to introduce more competition is a continuing issue in the public utility industry.

The most visible example of deregulation was the 1982 breakup of the Bell System. This actually started in 1974 as an antitrust suit by the U.S. Department of Justice. In the 1982 federal court settlement, AT&T agreed

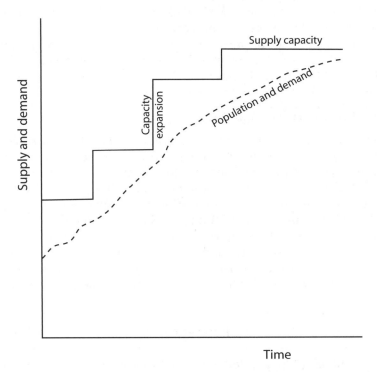

FIGURE 7-2. Demand and supply with capacity expansions

to divest itself of its local phone companies in exchange for the right to enter the computing business. AT&T retained its long-distance business, but it soon encountered intense competition from start-up companies. AT&T also lost Bell Labs, which became Lucent Corporation.

During the 1990s, electric power deregulation was attempted under state law in California. The main purpose of this deregulation was to unbundle integrated utilities and separate generation, transmission, and local distribution. A number of problems occurred, including bankruptcies and a spike in the cost of wholesale power. Electricity deregulation remains a controversial topic, with its supporters and opponents.

Researchers at the Competitive Enterprise Institute have estimated that on an overall basis, regulation—which extends well beyond utility regulation—cost the U.S. economy $1.13 trillion in 2005, some 10% of the U.S. gross domestic product (Miller 2006). Little can be said in defense of monopolies and against deregulation except that some utilities are usually not good candidates for competition. For example, water and wastewater utilities involve such heavy expenditures for fixed assets that competition would not normally be appropriate. Also, water is too heavy to transport efficiently in the same way that electric power can be moved across networks. However, you can move raw water by exchanging the right to divert it at different places. At the same time, convincing arguments can sometimes be made for the efficacy of privatizing water systems, as has been done in the United Kingdom and some other countries. Water privatization is by no means universally favored, however.

Utility Subsidies to Lower-Income Groups

Utilities play important social roles in providing essential public services to low-income people. Yet this also means that utilities face dilemmas when low-income families are unable to pay, and thus utilities have assistance programs and mechanisms such as "social rates" and "lifeline rates." The complex questions that arise in serving low-income people require a utility to know when a customer is really hard up or is trying to defraud them. There is no universal answer to this question, and thus utilities must walk a fine line between being businesslike and providing social services. Helping low-income customers involves socialistic approaches, and a patchwork of laws, subsidies, and charity operations is in effect to handle these situations.

References

About Inventors. (2007). Television history. http://inventors.about.com/. Accessed February 13, 2007.

Armstrong, E. L., ed. (1976). *History of public works in the United States.* Public Works Historical Society, American Public Works Association, Chicago.

Baltimore Gas & Electric. (2007). http://www.bge.com/. Accessed February 13, 2007.

City of Fort Collins, Light and Power Utility. (1985). *Partners in power: A history of the Light and Power Utility.* City of Fort Collins, CO.

Denver Post. (2006). "Cable TV bill has benefits and pitfalls." June 26.

Dow Jones. (2007). Dow Jones averages. http://www.djindexes.com/. Accessed August 12, 2007.

Edison Electric Institute. (2006). Key facts about the electric power industry. http://www.eei.org. Accessed August 11, 2007.

Grigg, N. (2006). "Water sector structure, size and demographics." *Journal of Water Resources Planning and Management* (ASCE), 133(1), 60–66.

Miller, H. (2006). "I have a dream: Scientific, logical regulation." *Wall Street Journal,* July 13.

Mushkin, S., ed. (1972). *Public prices for private goods.* Urban Institute, Washington, DC.

Natural Gas Supply Association. (2007). Industry and market structure. http://www.naturalgas.org. Accessed August 12, 2007.

U.S. Energy Information Administration. (2007). Energy basics 101. http://www.eia.doe.gov/basics/energybasics101.html. Accessed February 15, 2007.

Wasik, J. F. (2006). *The merchant of power: Sam Insull, Thomas Edison, and the creation of the modern metropolis.* Palgrave Macmillan, New York.

8

Construction Industry Economics

The Construction Industry in the United States

The construction industry is the largest sector within the national economic accounting framework and an important arena for infrastructure and environmental work. Construction economics deals with the industry's structure, economic policies, and conditions; construction spending; value added; the number of establishments; and the number of jobs. Though these topics deal with money, construction finance also extends to business arrangements, contracts, sources of financing, and financial regulation.

The chapter begins with the industry's structure and then explains the role and impact of construction in the economy, including key industry statistics. These include estimates of construction volume, value added to the economy, the number of construction firms and establishments, and construction industry employment. The chapter also identifies the construction sector's main players in order to give insight into its complexity and diversity. The roles of support groups, such as trade show organizers and publishers, are also explained in the chapter, which concludes with a brief discussion of the main laws that control the economics of the construction industry.

The Structure of the Construction Industry

The construction industry involves many design firms, regulators, and suppliers, as well as the constructors themselves. The "Career Guide to Industries:

Construction," published by the U.S. Bureau of Labor Statistics (2006), explains their roles. The industry has three major segments:

■ buildings contractors, who build residential, industrial, commercial, and other buildings;
■ heavy and civil engineering construction contractors, who build sewers, roads, highways, bridges, tunnels, and other projects; and
■ specialty trade contractors, who perform specialized activities related to construction such as carpentry, painting, plumbing, and electrical work.

General contractors coordinate construction. They usually specialize in one type of construction, such as residential or commercial building. They take responsibility for the complete job, and they may subcontract work to other contractors. Specialty trade contractors usually perform work in only one trade (such as painting, carpentry, or electrical work) or closely related trades (such as plumbing and heating).

The construction industry also includes regulators and support groups and suppliers. In the construction industry, the providers are some 7 million construction firms of varied types and sizes, along with another group of some 2 million "nonemployers" who are engaged in direct construction. Regulators are usually government employees engaged in ensuring health and safety, the quality of performance, and related areas. The support group for construction provides products and services including materials, equipment, design, consulting, and many others. Using these products and services along with labor, constructors assemble built systems that meet governmental codes and the standards set by private and public owners.

Figure 8-1 shows the structure of the construction industry: Constructors build the facilities for owners, both are supported by groups providing products and services, and regulators enforce a host of codes and standards. Figure 8-2 shows the iron triangle of the construction industry. Constructors and their direct supporters, such as consulting engineers, are at one corner of the triangle performing the direct work of the sector. Politicians and interest groups are at another corner of the triangle. These include groups working on issues of common interest that range from codes and standards to the availability of credit for construction to labor and other issues. Government agencies, with their elected and appointed officials, are the third corner of the triangle. These agencies control the policies and programs that affect the construction industry, such as finance, the stimulation of the economy, and the appropriation of funds. The appointed government officials administer programs, let contracts, and control flows of funds and programs that directly affect construction.

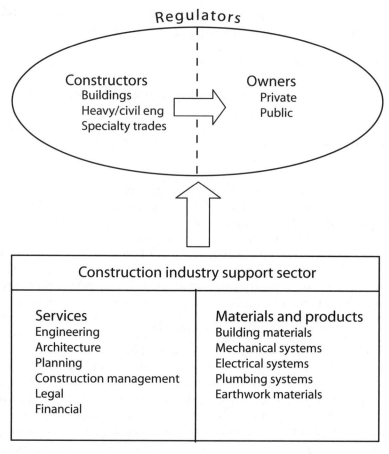

FIGURE 8-1. The structure of the construction industry

Construction's Role in the Economy

As the largest single industry within the categories of the North American Industrial Classification System (NAICS), construction plays a big role in the national economy. In the United States, it accounts for about $1 trillion in annual spending, stimulates businesses through the economic multiplier effect, sustains many jobs and business establishments, and offers an opportunity for government to jump-start economic development through investment and the stimulation of job creation. Of course, with the great financial crisis of 2008–9, construction spending fell sharply.

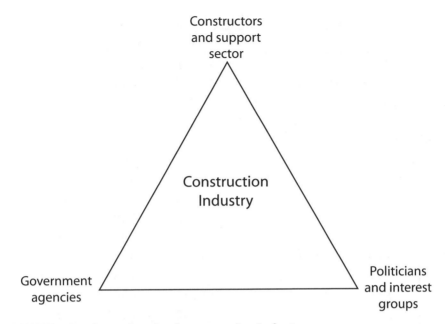

FIGURE 8-2. Iron triangle of construction industry

Construction Spending

The volume of construction activity is reported by the U.S. Bureau of the Census (2006a) as construction spending or construction put in place. This data source is reported in various places, including in *Engineering News-Record* (*ENR*) magazine's "Construction Economics" section. In 2005 private and public construction spending in the United States totaled about $1.1 trillion, with private construction contributing about 78%. Construction spending is not exactly the same as the gross domestic product (GDP) measure of the construction industry because some of the spending is accounted for in the GDP reports of other industries. That is why construction value added, as reported in Chapter 2, is closer to $600 billion, whereas construction spending is nearly double that figure. Construction spending is a measure of the total flow of funds to construction. National economic accounting provides another view of construction's impact through measurement of the value added to the economy's GDP.

Table 8-1 shows overall construction spending for 2005, distributed as private and public spending. Whereas Table 8-1 shows that almost all residential construction is private, Table 8-2 shows that of the total residential construction market, most construction is new, single-family homes. No

TABLE 8-1. Construction spending, 2005 (millions of $)

	Total	Private	Public
Total construction	1,120,649	874,077	246,571
Residential	633,395	626,815	6,580
Nonresidential	487,253	247,262	239,991
Lodging	12,033	11,790	
Office	46,420	35,277	11,143
Commercial	71,864	67,004	4,860
Health care	36,653	28,481	8,172
Educational	76,944	12,832	64,112
Religious	7,597	7,583	
Public safety	9,861	353	9,507
Amusement and recreation	19,020	8,031	10,989
Transportation	27,223	6,835	20,389
Communication	14,366	14,264	
Power	33,952	25,136	8,815
Highway and street	67,051		66,694
Sewage and waste disposal	17,842	267	17,576
Water supply	12,200	279	11,921
Conservation and development	5,364		5,232
Manufacturing	28,863	28,643	

Note: Within private construction, highway and street and conservation and development are included other categories. Within public construction, lodging, religious, communication, and manufacturing are included in other categories.

Source: U.S. Bureau of the Census 2006a.

wonder that in 2007, when the housing market was hit with falling demand and problems with subprime loans, it was of great concern to the whole economy. Of the private commercial and office construction markets, Table 8-2 shows that a big share is retail and office buildings, with automotive, warehouse, and farm construction trailing behind. Much of the construction spending for power projects is in the private sector, with distribution as shown by Table 8-3. Statistics give less detail on public spending, but it is interesting to compare the spending by level of government, as shown in Table 8-4. Most of the construction spending is by state and local governments, which are the main clients for civil engineers. The distribution of

TABLE 8-2. Private construction spending, 2005 (millions of $)

Type of construction	Amount spent
Residential	
New single-family	423,432
New multifamily	46,573
Improvements	156,810
Total	626,815
Office	
General	30,732
Financial	4,485
Total	35,277
Commercial	
Automotive	5,860
Food/beverage	7,750
Multi-retail	22,957
Other commercial (stores)	11,448
Warehouse	13,091
Farm	5,892
Total	67,004

Source: U.S. Bureau of the Census 2006a.

TABLE 8-3. Private power project construction, 2005 (millions of $)

Type of power	Amount spent
Electric	16,468
Gas	6,792
Oil	1,314
Total	25,136

Source: U.S. Bureau of the Census 2006a.

state and local spending (as shown Table 8-4) indicates that transportation, educational facilities, and water/wastewater utilities get the most funding.

The Number of Construction Firms and Establishments

The number of construction firms and establishments in the United States is reported by the U.S. Bureau of the Census (2006b), which defines an

TABLE 8-4. Public sector spending on construction, 2005 (millions of $)

Type of spending	Amount spent
Federal	17,690
State and local	228,881
Highways and streets	66,068
Educational	63,059
Transportation	18,701
Sewage and waste disposal	15,941
Water supply	11,721
Amusement and recreation	10,619
Public safety	8,654
Power	8,009
Office	7,664
Health care	7,351
Residential	5,049
Conservation and development	2,983
Commercial	2,698
Total federal state and local spending	246,571

Source: U.S. Bureau of the Census 2006a.

establishment as a relatively permanent office or other place of business. The Economic Census, which is taken every five years, provides these data for establishments in the United States across the various sectors and industries.

The 2002 Economic Census showed for the construction sector (NAICS 23) that there were 710,307 employer establishments (Table 8-5). Among these firms, 211,845 construct buildings, 49,826 work on heavy and civil engineering projects, and 448,663 are specialty trade contractors (U.S. Bureau of the Census 2006a).

In addition, the U.S. Bureau of the Census identified 2,071,317 nonemployer establishments in 2002. Nonemployers are defined as follows:

Most non-employers are self-employed individuals operating very small unincorporated businesses, which may or may not be the owner's principal source of income. Although non-employers constitute a large part of the business universe in terms of the number of establishments, they contribute a relatively small portion of the overall

TABLE 8-5. Establishments and employees in the construction industry, by size of establishment

Size of establishment (number of employees)	Number of establishments	Total employees
1–4	421,959	865,891
5–9	140,498	890,968
10–19	78,917	1,046,853
20–49	46,625	1,386,208
50–99	13,649	930,246
100–249	6,640	981,340
250–499	1,434	484,560
500–999	422	283,433
1,000 and up	163	323,571
Total	710,307	7,193,070

Sources: for the number of establishments, U.S. Bureau of the Census 2006b; for the number of total employees, U.S. Bureau of Labor Statistics 2007.

sales and receipts data. Tax-exempt businesses are excluded from the non-employers tabulations. (U.S. Bureau of the Census 2006b)

Nonemployer establishments generated receipts of $115 billion. Together, employer and nonemployer establishments in the construction industry accounted for $1.311 trillion of spending in 2002.

Construction Jobs

The approximately 700,000 construction firms provide about 7 million construction jobs, as shown in Table 8-5. Specifically, these 7,193,069 employees had a payroll of $254 billion; 5,317,758 of them were construction workers, and the rest held support jobs. The distribution of these jobs by occupation can also be obtained from Bureau of Labor Statistics data, as shown in Table 8-6. These occupations shown in this table account for about 99% of all construction industry jobs.

The largest segment of construction industry employment, which accounts for about two-thirds of the total, consists of workers engaged in "construction and extraction," or construction labor. Another large category is "installation, maintenance, and repair," at about 7%. Management is almost 5%, but when you add office and administrative and business and financial operations, you reach about 17% of all employees.

TABLE 8-6. Construction industry employment distribution

SOC	Occupation	All construction, Sector 23	Construction of buildings, Sector 236000	Heavy and civil engineering, Sector 237000	Specialty trade, Sector 238000
00	All	7,633,080	1,782,550	991,710	4,858,820
11	Management	362,950	151,730	57,530	153,690
13	Business and financial operations	215,190	75,370	28,260	111,570
17	Architecture and engineering	75,220	35,780	16,210	23,230
37	Building and grounds cleaning and maintenance	49,890	15,640	13,090	21,160
41	Sales and related	152,510	48,790	14,480	89,240
43	Office and administrative support	732,340	220,210	86,700	425,440
47	Construction and extraction	5,101,800	1,141,360	572,490	3,387,960
49	Installation, maintenance, and repair	531,060	34,870	93,480	402,710
51	Production	99,780	15,960	16,250	67,570
53	Transportation and material moving	280,150	29,670	82,530	167,950

Note: SOC is the Bureau of Labor Statistics' designation for "Standard Occupational Code."

Source: U.S. Bureau of Labor Statistics 2007.

Architecture and engineering represent only about 1% of construction industry employment. Of the 75,220 jobs in the architect and engineer category, some 24,140 were civil engineers and another 9,140 were civil and architectural drafters. Many civil engineers are engaged in the construction industry, but they are not counted as within it by the Bureau of Labor Statistics. They are found in the engineering services (consulting firms), government, and supplier categories. Of some 224,000 civil engineers in the workforce, about 195,000 work in these sectors (from Bureau of Labor Statistics data for 2004).[1]

Construction Costs

Construction spending of over $1 trillion annually depends on the costs of factor inputs, which include capital, labor, energy, materials, and services. Each of these fluctuates over time and in different places due to market conditions. ENR (2006) publishes indices of construction costs that enable us to track their variation with inflation.

ENR maintains 20-city national average indexes for construction costs (CCI), building costs (BCI), materials, skilled labor, and common labor. It publishes monthly prices on a rotating cycle. In a given month, the first week has prices for 21 products, including asphalt, cement, aggregates, concrete, brick, concrete block, and mason's lime. The second week, ENR includes prices for 20 pipe products covering reinforced concrete pipe, corrugated steel pipe, vitrified clay pipe, PE underdrain, polyvinyl chloride sewer and water pipe, ductile iron pipe, and copper water tubing. The third week features prices for 18 more products covering lumber, plywood, plyform, particle board, gypsum wallboard, and insulation. In the fourth week, ENR publishes prices for 16 more products covering structural steel, reinforcing bar, steel plate, metal lath, aluminum sheet, stainless steel sheet and plate, and H-piles.

The CCI and BCI are similar to the Consumer Price Index in that they compile multiple measures of cost into a single indicator. The BCI includes skilled labor (bricklayers, carpenters, and structural ironworkers), structural steel shapes, portland cement, and 2 × 4 lumber. Their labor index is the labor component of the BCI and it includes union wages and fringe benefits (carpenters, bricklayers, and iron workers). The materials cost index is the materials component of the BCI and includes structural steel, portland cement, and 2 × 4 lumber. The CCI includes common labor, structural steel shapes, portland cement, and 2 × 4 lumber.

---

[1]These data are from the interactive tables at the Web sites of the Bureau of Labor Statistics (http://www.bls.gov) and the Bureau of Economic Affairs (http://www.bea.gov).

The Global Construction Industry

ENR, using data from Global Insight, a management consulting firm, reported that global 2004 construction spending was $3.9 trillion. The current U.S. total, at some $1 trillion, is on the order of 25% of it. At any time, the relative states of the economies and their currencies will cause this percentage to vary, of course. Table 8-7 shows the distribution of construction spending.

Construction Industry Support Groups

As I explained earlier in the chapter, the construction industry's some 700,000 establishments are supported by other groups offering products and services as inputs to the construction process. This section provides some detail on these services and products.

"Services" is a broad term, and thus, according to the Bureau of Labor Statistics, it includes education and health, financial, information, profes-

TABLE 8-7. Distribution of global construction spending

Country	$ (billions)
United States	1,159.1
Japan	506.8
China	269.1
Germany	246.8
France	196.8
Italy	182.1
United Kingdom	177.5
Spain	165.9
Canada	123.3
Netherlands	78.5
India	73.9
Mexico	65.5
Brazil	54.3
Australia	49.3
Russia	42.1
All other	522.5
Total	3,913.5

Source: Tulacz 2005.

sional and business, equipment repair, and even government. Many of the groups included in these service categories identify their work as in the construction industry. For example, civil engineers and architects may consider themselves part of the construction industry, even if their primary participation is in preparing plans rather than engaging in construction.

Architectural and engineering services (NAICS 5413) in 2006 employed about 1.35 million of the total of some 17 million employed within the professional and business services category of NAICS Sector 54.

Within the engineering services sector (NAICS 54330), consulting engineering companies are a major segment. Table 8-8 shows the number of such establishments by size. These establishments are not all civil engineering firms. Actually, the Economic Census reports these data as "product lines," and the 2002 Census reported that 15,762 firms offered engineering services for transportation and 13,774 firms offered engineering services for municipal utilities. This gives at least an estimate of the number of firms in these markets. The U.S. Bureau of the Census (2006c) has developed a North American Product Classification System, and it has further data on the service industries, including engineering services.

The supplier sector of the construction industry is huge. A big share of the some $500 billion difference between value added and construction spending is for intermediate inputs in the form of mechanical,

TABLE 8-8. Number of consulting engineering establishments, by size, 2006

Employees	Establishments
0	271
1	8,374
2	5,650
3–4	6,120
5–6	4,110
7–9	4,208
10–14	4,258
15–19	2,436
20–49	4,813
50–99	1,509
100 and up	997
Total	42,746

Source: U.S. Bureau of Labor Statistics 2007.

plumbing, electrical, wood, brick, concrete, roofing, landscape, and other supply inputs.

Construction Industry Associations, Trade Shows, and Media

Given the size of the construction industry, it has formed a number of trade associations and associated shows. These are an important part of the structure of the construction industry and represent groups and places where business relationships can be created and influence can be brought to bear.

The Associated General Contractors of America (AGC) considers itself the voice of the construction industry. It has a national headquarters and state chapters. Trade shows for the construction industry attract many people and exhibits because the industry is so large. For example, a trade show called CONEXPO-CON/AGG 2008 expected 125,000 attendees for its 2008 Las Vegas convention. Its organizers include the Association of Equipment Manufacturers (main organizer); the National Ready Mixed Concrete Association; the National Stone, Sand & Gravel Association; and the AGC (CON-EXPO-CON/AGG 2007).

The construction industry also has a number of magazines and trade journals. Many of the industry's sectors or support groups publish a magazine or have a Web page. *ENR* magazine is focused on the construction industry, and as noted above, it devotes a special section of each issue to construction economics. The contents of this section illustrate main issues of interest to the industry.

Construction Laws and Regulators

The regulation of the construction industry occurs in every area where laws are written—health and safety, the environment, economic competition, business practices, and so on. For the construction industry, four categories are particularly important: health and safety, the quality of work, employment and labor, and the environment. For health and safety, the administration of the Occupational Safety and Health Act dominates regulation in the construction industry. The quality of work is regulated by codes and standards and the numerous service requirements of constructed facilities. Employment and labor are regulated by labor legislation such as the Davis-Bacon Act, minimum wage laws, and immigration laws. The Davis-Bacon Act, stemming back to the 1930s, sets wage requirements for all projects that receive federal assistance. Environmental rules also govern construction. Constructors must comply with sediment control, noise, wastewater control, and other environmental requirements.

References

CONEXPO-CON/AGG. (2007). http://www.conexpoconagg.com/index.asp. Accessed August 3, 2007.

Engineering News-Record (ENR). (2006). Construction economics. http://www.enr.com/features/conEco/subs/default.asp. Accessed May 18, 2006.

Tulacz, G. (2005). "World construction spending nears $4 trillion for 2004." *Engineering News-Record,* 254(1), 12–13.

U.S. Bureau of the Census. (2006a). Construction spending: Construction at a glance—2005 total. http://www.census.gov/const/www/c30index. html. Accessed May 8, 2006.

———. (2006b). Economic census, purpose, and use of non-employer statistics. http://www.census.gov/epcd/nonemployer/view/intro.html. Accessed May 17, 2006.

———. (2006c). North American Product Classification System. http://www.census.gov/eos/www/napcs/napcs.htm. Accessed May 18, 2006.

U.S. Bureau of Labor Statistics. (2006). Career guide to industries: Construction. http://www.bls.gov. Accessed May 17, 2006.

———. (2007). May 2006 national industry-specific occupational employment and wage estimates: Sector 23—Construction. http://www.bls.gov/oes/current/naics2_23.htm. Accessed July 6, 2007.

Part II

Finance for Infrastructure and the Environment

9

Infrastructure Finance

Financial Tools for Infrastructure Management

If engineering shaped infrastructure in the past, finance will be an important key to its future because past challenges to build complex systems have been overtaken by today's challenges to pay for them. If evidence of this is needed, we only need to point to the financial crisis, pressure on government budgets, and the need to make public services more efficient.

Chapter 1 of this book explained how economic and financial tools are used by a broad group of infrastructure and environmental managers who face different challenges than their predecessors. When planning for infrastructure, they can no longer simply "determine the need and build a system to meet it"—an old view of infrastructure planning and management. Instead, they now must apply demand management and innovation to meet needs according to "triple bottom line" goals in economic, social, and environmental areas.

The first part of the book addressed the economic issues and tools that are shaping this new management environment. This chapter introduces the book's second part, with its focus on financial principles. The book's third part gives the details of specific economic and financial tools for analyzing and managing infrastructure and the environment.

Managers of infrastructure systems must be able to "follow the money" to understand how their organizations work. It is a truism that if you do not understand the flow of money, you do not understand an organization. These managers will use financial experts for part of their work, and they may become de facto financial experts themselves. Most do not require mastery of the details of accounting, but they must generate the revenues needed and control the costs of their systems.

These infrastructure managers need knowledge of public and private sector financial principles, budgeting and financial planning, sources for capital and operating funds, debt financing, and cost control. In addition, they should be able to perform financial analysis and compute basic ratios, returns, and other financial indicators.

Like economics, finance has a private sector side and a public sector side (Fig. 9-1). On the private side, it addresses corporate finance, investments, and many other topics dealing with money, including personal finance. On the public side, it becomes the field of public finance, which applies its principles to public sector problems.

In the private sector, financial managers seek to maximize businesses' profit and shareholder value—that is, the return on their investment. This concept of maximizing return also applies to the public sector, but in a different way, because public managers must focus on how stakeholders value the outcomes of public investments and programs. Thus, it is harder to measure the return on investment for public programs than for private businesses because public outcomes involve all three legs of the "triple bottom line"—economic, social, and environmental.

Whether in the private or public sector, managers use finance to control costs, raise revenues, and prepare financial reports. To prepare them, a course or two in finance will be part of programs leading to master's in business administration or master's in public administration degrees. If graduates specialized in finance, their educations prepare them for jobs as business or government finance officers, bankers, stockbrokers, financial analysts, investment bankers, and similar roles.

The chapters ahead include a summary presentation of finance applied to the work of civil engineering, construction, and public works managers.

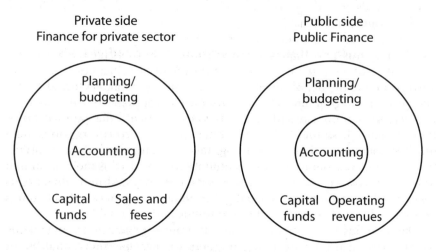

FIGURE 9-1. The public and private sides of infrastructure finance

The key elements of each topic are presented in enough detail to give infrastructure and environmental managers the essential tools they need.

The Evolution of the Field of Finance

Before about 1900, bookkeepers, bankers, businessmen, and other involved with money practiced finance, but it was not the organized discipline that it is today. Engineers had developed project finance, which became "engineering economics" (see Chapter 16). Economics had become a recognized discipline, and the field of finance split off from it about 1900 as business education became popular.

Finance began to emerge as a distinct field when big companies and trusts such as Standard Oil and U.S. Steel became monopolies and required new controls. Utilities also became monopolies, because business barons saw them as attractive opportunities for profit. Antitrust programs dissolved these giant businesses and opened up stocks and bonds as investments for individuals. In the 1930s, the Securities Act (1931) and the Securities Exchange Act (1934) increased the regulation of business, including financial reporting. Concepts such as capital budgeting and cash and inventory management became common financial management tools.

At about the same time, the public sector began to increase its scrutiny of financial management. Municipal budgeting and reporting reduced corruption and made local government more transparent. At the federal level, the Bureau of Budget was established as the predecessor of today's Office of Management and Budget, which is part of the Executive Office of the President and prepares the U.S. budget. Today, public finance is well organized at all three levels of government.

Finance for Business, Government, and Individuals

The lessons you learn about business finance also apply to personal affairs, and vice versa. Business finance covers issues concerned with financial markets, institutions, securities, risk management, and many other topics. Public finance is similar to business finance, but it also covers government functions such as political budgeting, raising public funds, accountability to taxpayers, and public sector accountability. Personal finance is similar to business finance, but it involves your own money and individual affairs such as balancing a checkbook. Table 9-1 shows how you can readily transfer the concepts of business finance to public and personal finance.

Regardless of the arena, the central issues of finance are organizing its functions, planning and budgeting, raising revenues and capital funds, planning and managing expenditures, accounting, and reporting. Financial

TABLE 9-1. Comparable concepts of business, public, and personal finance

Business finance	Public finance	Personal finance
Sales	Tax and fee revenue	Income
Business budget	Agency budget	Personal budget
Accounting	Public accounting	Personal records
Property	Property	Property
Liability and other insurance	Liability and other insurance	Personal insurance
Business borrowing	Government borrowing	Personal credit
Paying taxes	Collecting taxes	Paying taxes
Capital budget	Capital budget	Buying major items
Employee benefits	Public retirement programs	Retirement account
Corporate philanthropy	Managing philanthropy	Personal charity

managers control current and fixed financial assets to balance the enterprise's risks and profitability. This means that they must monitor revenues and costs, arrange credit, manage debt, manage budgets and funds, recommend dividends, and perform many other tasks.

The organization of financial functions varies from one place to another. In a business, the person who handles finance might be a chief financial officer, a vice president for finance, a controller (comptroller), or simply a business officer. In a public agency, there might be a director for finance and administration. Regardless of the place, the names on the door might include budgeting, accounting, auditing, assessments, purchasing, and treasury. Figure 9-2 illustrates the functions of a typical public finance office.

Budget processes are of critical importance to public infrastructure agencies. The budget office compiles and presents a consolidated budget to the chief executive officer, who presents it to the governing board or approval authority. Chapter 11 explains the budget process as it applies to any type of organization.

The accounting section of an organization is the basic building block for financial management. It maintains the records and prepares the reports that are the basis for making decisions. Internal or external auditors coordinate with the accounting office to make independent assessments of the organization's financial condition.

In public organizations, property tax collection is one of the main sources of revenue. Chapter 12 explains how property taxes are computed. The assessment office manages the accounting required for property assessment. This is usually a function of a county government.

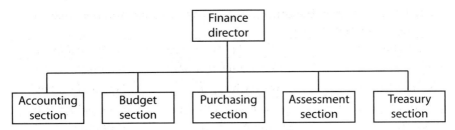

FIGURE 9-2. The divisions of a typical public finance office

Much of the financial control function is handled by the purchasing office, which, along with the accounting office, oversees the physical assets of an organization, makes sure that the organization gets good value, and sees that property is tracked and managed.

Another financial area is the treasury office, which makes collections and writes checks.

Financial Topics for Infrastructure and Environmental Managers

Accounting

Accounting is the language of business. Accountants keep the records of an organization's business transactions and prepare reports of net income and the balance among assets and liabilities. The term "accounting" is also used loosely in technical organizations to refer to such management functions as performance accounting, water accounting, or any process of counting and measuring something. Chapter 10 explains the basic principles of accounting, including the basic rules of private and public accounting and related functions, such as the financial audit.

Public Finance and Budgeting

Public finance, which is explained in Chapter 11, is the field that controls the management of public funds for purposes such as infrastructure and regulating the environment. Public finance is distinct from private finance because controls on funds are different. Governments are more subject to self-regulation than are private businesses, with authority vested in the political process and civil and criminal statutes. In the private sector, businesses are regulated by the Internal Revenue Service, the Securities and Exchange Commission, and other federal, state, and local entities.

Although budgeting is a critical financial tool in all organizations, it is especially important in the public's business because of the transparency needed to open government functions to public view and give the public a say in how government runs. How this works is explained in Chapter 12.

The budget is more than just a tool for allocating money. It requires decisions about the policies and directions of the organization, including the level of taxation and charges to the community or the fraction of total community resources needed for governmental programs and services. In newer approaches to budgeting, this is referred to as the "cost of government" (see Chapter 17). Management tasks revolve around the budget, including capital and operations planning. Budgets are structured to follow the programs and divisions of an organization. Before the budget is adopted, it is the proposed budget. After it is authorized, it becomes the official plan for the fiscal year.

Utility and Public Works Finance

Chapter 12 outlines sources of revenues for the industries that are considered utilities or public works services. It includes a diagram to illustrate the general flows of utility finance, to include capital and operational funding. Public works and utility finance are very important in the planning and operation of infrastructure and environmental programs.

Capital Markets

The sources of capital financing for building and rebuilding infrastructure systems are explained in Chapter 13, which covers how the capital markets operate. The topics discussed there include planning and justifying capital spending, capital budgeting, financing system expansion and renewal, sources of capital financing, stock and bond markets, capital market regulation, credit and risk, and development banks.

Asset Management

Chapter 14 explains how the principles of management and finance are being applied to the new field of asset management as it concerns infrastructure systems. In a general sense, asset management refers to any system to manage assets in order to sustain them and get a good return from them. Infrastructure asset management is especially important because the systems are so capital intensive.

Decision and Institutional Analysis

Financial management involves many decisions that require valid and reliable decision information. This information must include intelligence on institutional issues as well as pure numerical data. Chapter 15 covers the use of decision information and how to factor institutional analysis into decisions. Institutional analysis is a practical discipline that explains how the world really works and goes beyond what shows up on the books. For example, for financial situations, it will probe the stability of an organization, whether it is able to repay, and how sensitive it will be to shocks, among other questions.

Engineering Economics, Financial Analysis, and Planning

Infrastructure and environmental managers often use quantitative methods for financial analysis. These range from computing a loan payback and basic cash flow analyses to complex feasibility studies. Chapter 16 presents these basic quantitative methods, which are also referred to as engineering economics. Both engineering economics and financial analysis focus on quantitative analysis techniques, but they differ in that engineering economics is a niche field meant to focus on engineering problems, along with their broad social and environmental objectives.

As outlined in Chapter 16, financial analysis includes revenue analysis, cost analysis, institutional analysis, ability-to-pay analysis, secondary impacts analysis, and sensitivity analysis. These techniques are used in personal, business, and public activity, and they offer an effective framework to organize a financial or business plan. A financial plan has common elements across the platforms of business, public, and personal finance. Table 9-2 compares these planning activities.

Financial Strategy and Trends

The field of finance includes many lines of strategy that are meant to manage risk or exploit opportunities to improve profitability. One strategy is known as "arbitrage," which means borrowing money one place and then investing it somewhere else to make a profit. For example, the entrepreneurial manager of a utility might see that a subsidized fund for utility borrowing has created an opportunity to raise capital at a low rate, whereas he could turn around and invest it in short-term instruments at a higher rate. He could then borrow the maximum from the low-rate source and make a profit for the utility, thereby holding down rates for customers and increasing the utility's reserves and reducing its risk. The only problem is that this attractive strategy is made possible only by exploiting the public's resources in the subsidized fund. Therefore, this kind of arbitrage involving public funds will normally be prohibited by law. Arbitrage is, however, a regular strategy of private financial investment firms seeking to make profits for their customers.

TABLE 9-2. Comparison of activities among business, public, and personal finance

Planning activity	Business finance	Public finance	Personal finance
Revenue analysis	Sales	Taxes and fees	Income analysis
Cost analysis	Business costs	Government expenditures	Personal expenses
Institutional analysis	Business track record	Public institutions	People involved, reliability
Ability-to-pay analysis	Track record, business assessment	Cash flow analysis	Payments record, overall assessment
Sensitivity analysis	Reliability of forecasts	Reliability of forecasts	Reliability of forecasts
Secondary impacts analysis	Effects on other businesses	External impacts of activities	Other people involved

Debt management is another financial strategy with broad implications for infrastructure managers. When inflation and interest rates are considered over the short and long terms, an analysis might show that borrowing should be increased or decreased to minimize the public's long-term expense of using infrastructure facilities. Sensing the trends in inflation and interest rates is tricky and requires expert advice as well as a good bit of luck.

In private sector businesses, strategy sometimes revolves around detecting value or opportunities that are underrepresented in a stock's value. For example, a corporate raider might see that a large business with several subsidiaries would be worth more broken up than it is when combined. Therefore, the raider might seek to buy a controlling interest of the stock, force a breakup, then sell the separate equities for a profit. There are, of course, many other areas of financial strategy, and they apply to businesses operating in the infrastructure and environmental fields, just as they do to other industrial sectors.

Engineers who read this book may wonder about the term "financial engineering," which is used widely today in finance. It does not refer to the traditional practice of engineering but instead refers to the design of financial instruments, such as derivatives, to improve the performance of financial portfolios. With these, you can hedge risk, diversify, and take other steps to structure your finances. Although the term includes the word "engineering," it is really an application of analysis techniques to finance, rather than a form of "engineering," as most engineers understand the term.

Society is litigious, and the legal system is used often to resolve disputes as well as to control businesses and agencies. Financial risk is inherent in infrastructure and environmental management scenarios, and as a result, business law and financial regulation are major issues for civil engineering, construction, and public works managers.

Risk in an infrastructure or environmental organization covers multiple threats that range from natural disasters to employee lawsuits. Regardless of the type of risk, the financial implications of mitigating it will be important and include decisions such as whether to be self-insured, the level of financial reserves to maintain, whether to invest in security measures and mitigation systems, and the degree of internal financial controls to have. Whereas a financial manager in a public company is scrutinized by internal auditors, the Internal Revenue Service, and the Securities and Exchange Commission, the manager of a public infrastructure enterprise might be surprised by disclosures that go unnoticed in a more politically regulated business environment.

Applications to Infrastructure and the Environment

By applying financial principles to the management of infrastructure, managers can "follow the money" through budget requests, appropriations, expenditures, and the reporting of results. These principles evolved from roots in banking and investments to today's complex field of finance, which includes public and private finance.

The managers I have surveyed thought finance is very useful and can be learned on the job better than some other topics. One reason that it can be learned on the job is that it is so useful and has common elements across business, public, and personal finance. If an infrastructure manager has learned the principles and elements of finance in a small business or even through personal finance, many of them can be transferred to the arena of public finance. The knowledge that needs to be added is mostly about how public financing systems work.

The next few chapters outline the financial principles of accounting, regulatory controls, capital markets, and how rates and fees are set. After that, several chapters present specific tools that draw substance from finance, including asset management, decision analysis, and financial analysis itself.

10

Accounting and Financial Statements

Accounting: The Language of Business

As the language of business, financial accounting is essential in both private and public sector infrastructure work. Accountants provide the financial information and analysis that managers need to make decisions and to evaluate how their organizations and assets perform. Managers work regularly with a branch of accounting known as management accounting, which provides the information they need to make decisions and control their organizations.

Accounting tasks range from maintaining personal records to complex accounting for businesses and government organizations. They cover the methods used to track income and expenditures and to maintain financial records and reports. Few infrastructure and environmental managers study accounting before they face financial decisionmaking, and fewer still have studied public sector accounting. Thus, it is essential that they learn the basics of accounting.

Accounting in both the private and public sectors follows a core set of Generally Accepted Accounting Principles (GAAP), which are officially mandated for the accounting profession. Although infrastructure organizations mainly use government accounting practices, their accounting tasks are still similar to those of other organizations, requiring an analysis of labor, materials, and other expenses.

Most infrastructure organizations have a heavy percentage of fixed assets, which government accounting has not handled well in the past. This is an important issue because without good records, it is hard to make a case for infrastructure maintenance and renewal. There is cause for optimism, however, because new accounting rules require more information about fixed assets than in the past (see Chapter 14).

This chapter sets forth the main issues of accounting for private and public sector organizations. It provides an introduction to the main accounting reports used in infrastructure and environmental decisions, and it illustrates how to "follow the money" so that managers can understand their operations. It includes a section on "triple bottom line" accounting, which enables organizations to report social, environmental, and economic results, along with their financials.

Principles of Financial Accounting

The Structure of Accounting

Financial accounting has a logical structure to track and record the "stocks and flows" of money. "Stocks" are inventories of any quantity, such as money, fuel, water, or other commodities. "Flows" are rates of change.

In basic accounting, transactions are recorded with bookkeeping techniques and formats, which make possible the preparation of financial statements and reports. These financial records support the decisions of managers, boards of directors, customers, and regulatory agencies.

At the basic level, accounting information includes time sheets, vehicle logs, receipts, and myriad other reports of daily operations. The results of these are aggregated to become the reports upon which decisions about an organization's financial control are based. Information from the organization's books is needed to create checks and balances for the control of purchasing, fixed assets records, inventory, and hiring. The annual audit then constitutes an independent check on the organization's financial health.

Accounting for Different Types of Organizations

Accounting differs for private companies, public companies, utilities regulated by public agencies, nonprofit organizations, and government organizations. These organizations are distinguished by the following attributes:

- *Private company:* In private ownership but has not issued stock to the public. Example: a consulting engineering company.
- *Public companies:* Can raise capital by selling stock to members of the public, who become owners. These firms must follow rules of disclosure for financial information. Example: General Electric Corporation.
- *Regulated utilities:* These may be either private sector utilities or public companies. They are subject to the rules of state public utility commissions. Example: Pacific Gas and Electric Corporation, California.
- *Private sector nonprofit organizations:* These include public interest organizations, such as those that are tax exempt under Section 501(c)3 of the

Internal Revenue Code. They must comply with state rules and the rules of the Internal Revenue Service. Example: Road User Federation.

■ *Government organizations:* These include government agencies at the federal, state, and local levels. They must comply with the rules of government accounting, as well as with the GAAP. Example: East Bay Municipal Utility District, Oakland.

Processes of Accounting

When transactions occur, they are recorded in journals, from which they are posted to ledgers. This posting of transactions occurs in a system of "double-entry accounting," which tracks debits and credits as they affect assets, liabilities, and/or equity.

Bookkeeping, the mechanical part of accounting, shows how transactions become debits or credits to different accounts. Accounting entries can be posted under the cash or accrual basis of accounting, meaning when cash changes hands or when transactions occur.

Accounting Standards

The purposes of accounting standards are to make financial procedures and reports credible, concise, transparent, understandable, and comparable. The GAAP are managed by the accounting profession through the Financial Accounting Standards Board (FASB). The Government Accounting Standards Board (GASB) was established in 1984 as a companion to the FASB to augment the GAAP with specific standards for government organizations. The GASB sets accounting standards for state and local governments. Both the FASB and GASB are overseen by the nonprofit, tax-exempt Financial Accounting Foundation (FAF).

Before the inception of the GASB, a National Council on Government Accounting set guidelines as a committee of the Government Finance Officers Association of the United States and Canada (GFOA), which was formerly known as the Municipal Finance Officers Association. Today, the GFOA acts in an advisory capacity to the GASB. Infrastructure and environmental managers working in the public sector can benefit by learning about the GASB's principles for public sector accounting.

Since 1973, the FASB has been the main private sector authority to establish standards for private sector financial accounting. The FASB's standards are recognized by the Securities and Exchange Commission (SEC) and the American Institute of Certified Public Accountants (AICPA). The SEC has authority over reporting standards for publicly held companies under the Securities Exchange Act of 1934, but it relies on the FASB to set standards in the public interest.

The FASB is independent of other business and professional organizations. Before 1973, accounting standards were established by the AICPA. Today, the FASB consults on technical issues with the Financial Accounting Standards Advisory Council (FASAC),which has more than 30 members who prepare, audit, and use financial information. The members of the FASB and the FASAC are selected by the FAF.

The FAF's trustees come from constituent organizations with an interest in financial reporting. The constituent organizations make nominations, which are approved by the trustees. The trustees also select other trustees at large. The constituent organizations that participate in the FAF are the American Accounting Association; the AICPA; the CFA Institute; Financial Executives International; the GFOA; the Institute of Management Accountants; the National Association of State Auditors, Comptrollers, and Treasurers; and the Securities Industry and Financial Markets Association. Figure 10-1 shows how the authority to set financial standards is distributed.

The FASB has an extensive and open due process system that is modeled on the Federal Administrative Procedure Act. After an issue has been through

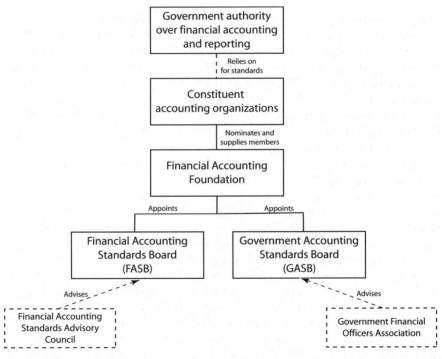

FIGURE 10-1. How authority to set financial standard is distributed

the process, the result is likely to be a Statement of Financial Accounting Standards. This statement will explain the standards, the effective date and transition, background information, research done, and the basis for the decision. It also identifies the voters for and against and their reasons, like a court decision. In addition to these statements, the FASB also issues Statements of Concepts that set forth accounting principles (FASB 2007).

Given the trends toward globalization, accounting standards should become more uniform across countries. Within the European Union, for example, the financial framework is seeking uniformity in accounting, and global trade requires more uniform standards worldwide. The International Accounting Standards Board in London promulgates standards for consideration by national governments. The United States has begun to experiment with use of international standards, but some time will be needed to consider a full shift toward their use.

Financial Reports

Financial Documents and Statements

Infrastructure and environmental managers normally begin to see financial documents in the form of the budget, which is a financial plan (covered in Chapter 11). Later, they might see an appropriation ordinance or law, and as they are implementing the budget, they will see financial statements that report on results after the budget is implemented. The two basic financial statements are the income statement (statement of revenues, expenses, and changes in retained earnings) and the balance sheet. A statement of change in financial position is also used, and cash flow statements are useful for many purposes.

The Income Statement

Income statements report the differences between revenues and expenditures over a period of time, such as a month or a year. Usually, the report is for the interval of a fiscal year. For engineers and hydrologists, the income statement can be compared with an annual water budget, for which the report is of inflows, outflows, and change in storage. In a business this is also called the profit-and-loss statement, and it reports how well the enterprise did during the last accounting period. For example, in personal finance or in a small business, the income statement shows whether there were profits or surpluses in a year, or whether the year ended in a deficit.

To illustrate an income statement and the basic issues of accounting, here I use the example of a small, private sector consulting engineering company. (Government accounting is in some ways simpler because it lacks the tax items that are found in private sector accounts.) This company, Rocky Mountain Infrastructure Consultants (RMIC), is a fictitious five-person "niche" consulting firm that does specialty work, in this case modeling and software development and sales. RMIC's income statement for the fiscal year ended December 31, 2006, is shown in Fig. 10-2. The statement shows that during 2006, RMIC's fees for services and sales of software were $500,000, or $100,000 per employee. The firm paid $50,000 in fees to various suppliers that helped it create software. All its operations expenses (salaries, rent, utilities, etc.) totaled $300,000, and the depreciation on its equipment was $30,000, leaving it with an operating profit of $120,000.

Fifteen years ago, RMIC borrowed $700,000 in capital to start up (the balance on this loan has now been reduced to $200,000), and its interest charge for the year was $16,000. It had to pay federal and state corporate income tax on its earnings before taxes, and these amounted to $20,000. Thus, its

Rocky Mountain Infrastructure Consultants Income Statement ($ dollars) Year Ended December 31, 2006	
Fees and sales	500,000
Cost of software sold	50,000
Gross profit	450,000
Operations expenses	300,000
Depreciation	30,000
Operating profit (earnings before interest and taxes)	120,000
Interest expense	16,000
Earnings before taxes	104,000
Taxes	20,000
Earnings after taxes	84,000
Preferred stock dividends	30,000
Earnings to shareholders	54,000
Common stock shares outstanding	10,000
Earnings per share	5.40

FIGURE 10-2. Income statement for Rocky Mountain Infrastructure Consultants

Source: Adapted from Block and Hirt 1997.

earnings after taxes were $84,000. As preferred stockholders, RMIC's five staff members paid themselves dividends of $30,000, thus leaving $54,000 that could be paid as dividends to holders of common stock or left in the business as cash for investment. Note that preferred stock and common stock are classes of ownership, as explained in Chapter 16. Because there are 10,000 shares of common stock outstanding, earnings per share were $5.40.

In Fig. 10-2, notice that the income statement's entries are flows of money, such as income and outgo. The balance statement provides a better picture of the stocks of money.

The depreciation shown on the income statement is not an actual cost until the money is spent. If the corporation's capital is not replenished by investing the depreciation, then the money remains in its bank account, but its other assets decline in value.

The Balance Sheet

Information about assets and liabilities is reported on the balance sheet or statement of financial position. Whereas an income statement is like a report of water flow and additions or deductions from a reservoir, the balance sheet is like a snapshot of how much water is in the reservoir at the end of the year, along with how much is owed to users and how much is expected from others at that point in time.

The balance sheet provides a report of changes in assets and liabilities over the accounting period. It is a cumulative snapshot of the financial picture of an accounting unit, whether an individual, a business, or a government agency. The balance sheet shows how the finances balance at an instant in time. It reaches a "balance" by including "earned surplus" or "net worth" in a statement.

On the balance sheet, the difference between assets and liabilities is "equity" or "capital," which provides the basic accounting equation:

$$\text{Assets} = \text{Liabilities} + \text{Equity}$$

Equity is like net worth or the accumulated profits, losses, investments, and other changes in the firm's financial position.

RMIC's balance sheet is shown in Fig. 10-3. RMIC was created 15 years ago (January 1, 1992), and now has a financial track record, which is reflected in its balance sheet. RMIC has $100,000 in cash on hand (in a savings/checking account), $200,000 in accounts receivable (which are reduced by $50,000 to account for bad debts its staff think they will never receive), an inventory of $300,000 (i.e., software ready to sell), and $50,000 in prepaid expenses (advanced to a software supplier to assist in financing a new product). Notice that these are all assets, but some are more liquid than others.

Rocky Mountain Infrastructure Consultants
Balance Sheet ($)
December 31, 2006

Assets

Current assets	
Cash	100,000
Accounts receivable	200,000
Less allowance for bad debts	50,000
Net accounts receivable	150,000
Inventory	300,000
Prepaid expenses	50,000
Marketable securities	0
Total current assets	600,000
Other assets: investments	50,000
Fixed assets less accumulated depreciation	450,000
Total assets	1,100,000

Liabilities and stockholders equity

Current liabilities	
Accounts payable	250,000
Notes payable	200,000
Accrued expenses	50,000
Total current liabilities	500,000
Long term liabilities	0
Total liabilities	500,000
Stockholders equity	
Preferred stock, 1,000 shares, $100 par value	100,000
Common stock, 10,000 shares, $1 par value	10,000
Retained earnings	490,000
Total stockholders equity	600,000
Total liabilities plus stockholders equity	1,100,000

FIGURE 10-3. Balance sheet for Rocky Mountain Infrastructure
Consultants

Source: Adapted from Block and Hirt 1997.

The firm does not have any marketable securities, but it could choose to invest some of its cash in them. Summing these yields current assets of $600,000. The firm also has long-term investments of $50,000 (as opposed to liquid, current investments) and net fixed assets of $450,000 (vehicles, computers, furniture, etc.). Adding all these yields results in total assets of $1,100,000. Note that the special category of fixed assets is important in infrastructure analysis, and it will be discussed in detail later in the chapter. The balance sheet could have shown fixed assets by their original cost and also showed the depreciation to deduct.

Liabilities and stockholders' equity begin with current liabilities in the form of accounts payable, notes payable, and accrued expenses, yielding total current liabilities of $500,000. There are no long-term liabilities (such as a long-term loan that requires no current payments). Thus, total liabilities are $500,000.

Stockholders' equity begins with the total par value of the preferred and common stock, which is $110,000. Retained earnings are computed as the number that satisfies the accounting equation of assets = liabilities + equity. Thus, retained earnings are computed as $490,000. Together with the par value of stock, this makes the category of stockholders' equity come to $600,000 which, when added to total liabilities, makes liabilities and equity come to $1,100,000, which "balances" with the total assets. This is no surprise, because retained earnings were computed to make this balance happen. In other words, the variable is retained earnings, which is adjusted to balance assets with liabilities plus net worth.

Statement of Retained Earnings

A statement of retained earnings is sometimes used to show changes over a year. Figure 10-4 shows the statement of retained earnings for RIMC.

Rocky Mountain Infrastructure Consultants Statement of Retained Earnings ($) Year Ended December 31, 2006	
Retained earnings, balance on January 1, 2006	436,000
Add: Earnings available to common stockholders, 2006	54,000
Deduct: Cash dividends declared for common stockholders, 2006	0
Retained earnings, balance on December 31, 2006	490,000

FIGURE 10-4. Statement of retained earnings for Rocky Mountain Infrastructure Consultants

Statement of Cash Flow

Since 1987, a statement of cash flows has been required by the FASB (Block and Hirt 1997). This statement overcomes some of the difficulty in interpreting the balance sheet, which arises due to depreciation accounting and other issues that tend to obscure actual performance. The difficulty in accounting for depreciation leads to disparities between the market value and book value of equities. Market value means how much a company would actually be worth if it were to be sold, whereas its book value records how much it cost, how much it owes and is owed, and similar information that is based on actual transactions.

The statement of cash flows has three parts: cash flows from operating activities, cash flows from investing activities, and cash flows from financing activities. The simplest method to prepare this statement, called the "indirect method," starts with net income. Figure 10-5 shows the statement of cash flows for RMIC.

Some of the entries in the cash flow statement require you to look at the past year's balance sheet to see the changes, but you can understand the concept from this explanation. Changes in cash flows focus on the organization's cash position from money on hand or in the bank. Net income represents cash changes during the year as reported on the income statement, subject to adjustments of items on the income statement and the balance sheet that represent accounting values but not necessarily cash values. In this case, you see a $60,000 increase in cash from operations as a result of adjustments for depreciation and the other items shown. You see a $40,000 increase in cash from investment activities and a $30,000 decrease in cash due to preferred stock dividends paid. This sums to a $154,000 increase in cash for RMIC for 2006. This amount is not shown on the other statements. The entries in the cash flow statement require further explanation, but the details have been omitted.

Pro forma financial statements can be prepared to project results expected in future periods. These are useful in financial planning and in sensitivity analysis, which examines the variable sensitivity of difference factors under alternative "what if" scenarios for financial strategy.

Annual Reports

The annual report is a useful device to report financial results and to summarize the accomplishments of an organization. Financial statements are found in the annual reports for any public company and for many government agencies and enterprises. They normally include descriptive, financial, and statistical information.

The annual report focuses management's attention on results achieved and financial health. In public companies, the primary goal is the "bottom

Rocky Mountain Infrastructure Consultants Statement of Cash Flows ($) Year Ended December 31, 2006		
Net income (earnings after taxes)		$84,000
Adjustments to determine cash flows from operations		
Add back depreciation	$30,000	
Increase in accounts receivable	(40,000)	
Increase in inventory	20,000	
Decrease in prepaid expenses	10,000	
Increase in accounts payable	50,000	
Decrease in accrued expenses	(10,000)	
Total adjustments		60,000
Cash flows from investing activities		
Increase in long-term investments	0	
Increase in plant and equipment	40,000	
Net cash flows from investing activities		40,000
Cash flows from financing activities		
Increase in bonds payable	0	
Preferred stock dividends paid	(30,000)	
Common stock dividends paid	0	
Net cash flows from financing activities		(30,000)
Net increase or (decrease) in cash flows		$154,000

FIGURE 10-5. Statement of cash flow for Rocky Mountain Infrastructure Consultants

line," with stockholders holding management accountable for profits and the appreciation of stock. In government agencies or enterprises, the approach is to present results more objectively without a focus on financial profit but on benefits for the citizenry as a whole. This is what is measured by "triple bottom line" reporting, which I describe later in this chapter.

The financial report contains the results of the past year's activities, including operational and fiscal performance. Management decides how to present financial statements, in compliance with financial standards.

The requirement for an annual financial report for public companies is specified by SEC regulations. The GASB requires financial statements, and government agencies are required to prepare comprehensive annual financial reports. The contents of the other elements of an annual report (descriptive and statistical information) are not specified by regulatory authorities. In government organizations, the annual report is recognized as an important part of management's report to citizens, and popular reports may also be prepared without any set format.

Reforming Financial Reporting

It should be clear that the potential for abuse in financial reporting is high. In the private sector, companies and individuals have been accused of "cooking the books" to create false results so they can deceive someone. For example, if a chief executive is under pressure to have continuous growth in profits, he might pressure the financial officer to reduce depreciation or inflate the value of receivables to make the company look more profitable. These deceptions would be hard to detect without in-depth checks.

As a result of high-profile scandals in the early 2000s, Congress passed the Sarbanes-Oxley corporate governance law, which in 2002 created the Public Company Accounting Oversight Board. The SEC selects the board's five members and oversees it (Reilly 2006). This has been controversial, in that it created a lot more paperwork and thus required more investment in accounting, which was a boon for accounting firms. Some say it has helped improve corporate accountability; others say it is nothing but another drag on U.S. competitiveness.

Financial and Performance Auditing

Audits are required to assure the integrity of an organization's financial statements. A financial audit is the process of examining accounts or making an outside check on the validity of the financial management and the health of the enterprise. As a regulatory measure, an audit is generally carried out by different accountants than those who create the regular accounts. The principle is the same as having one engineer check the work of another one.

In the public context, the U.S. Government Accountability Office (GAO) uses the term "performance audit" to extend financial audits into performance evaluation, which includes financial, economic, and programmatic audits. The GAO (1982) defines these as:

■ Financial compliance determines (a) whether financial operations are properly conducted, (b) whether the financial reports of an audited entity are presented fairly, and (c) if the entity has complied with applicable laws and regulations.

■ Economic efficiency determines whether the entity is managing or utilizing its resources (personnel, property, space, and so forth) in an economical and efficient manner and the causes of any inefficient or uneconomical practices, including inadequacies in management information systems, administrative procedures, or organizational structures.

■ Program results determine whether the desired results or benefits are being achieved, whether the objectives established by the legislature or other authorizing body are being met, and whether the agency has considered alternatives that might yield the desired results at a lower cost.

Government Accounting

Although government accounting also follows the GAAP, its requirements differ in some ways from those for private firms. For one thing, no reports to the SEC are required, and government books face scrutiny from elected boards and the public rather than corporate boards.

The GASB sets the basic standards for state and local governments. The federal government uses accounting specified by the FASB. To illustrate how the GASB sets the rules for state and local government accounting, the following principles were developed by the GASB and adopted for municipal governments by the National Council on Governmental Accounting (Miller and Warren 1991):

1. A government accounting system must fully disclose the operations of the government unit and comply with the GAAP and legal and contractual provisions.
2. Government accounting systems should be organized around "funds," which are segregated to focus on special program operations.
3. Types of funds include the general fund, special revenue funds, capital project funds, debt service funds, enterprise funds, and internal service funds.
4. The number of funds should be minimized to eliminate undue complexity.
5. Accounts should distinguish between fixed assets and long-term liabilities in proprietary funds (enterprise funds) versus other funds.
6. Fixed assets should be accounted for at cost (i.e., historical cost).
7. The depreciation of fixed assets "should not be recorded in the accounts of governmental funds. Depreciation may be recorded in cost accounting systems or calculated for cost funding analyses, and accumulated depreciation may be recorded in the general fixed asset account group." (On this, see the discussion below of "GASB 34.")
8. The modified accrual or accrual basis of accounting should be used. This means that expenses and revenues are credited when they occur, rather than when the cash is received or disbursed, so the financial reports will reflect a picture of the entity's actual financial health at all times.

9. An annual budget should be used by every government unit.
10. Interfund transfers and proceeds of long-term debt should be recorded separately from fund revenues and expenses.
11. Common terminology should be used in budgets, accounts, and financial reports.
12. Financial reports should be prepared to facilitate management control, legislative oversight, and external reporting.

Principles 5, 6, and 7 are related to fixed assets, but they were modified by GASB 34, which is explained below.

Fund Accounting

Fund accounting (see GASB Principles 2 though 4 above) is an important feature of accounting for both public sector and nonprofit activities. It is used to show accountability rather than profit or loss. Funds are self-balancing accounts to report expenditures by designated purposes. Whereas for-profit businesses might only have one set of balancing accounts or a general ledger, government agencies and nonprofits can have more than one general ledger. Funds are designed by account numbers.

The National Council on Government Accounting suggests these categories of funds:

■ *General fund:* To account for most of a municipality's operations, such as general administration and police.
■ *Special revenue funds:* Used for specific purposes, such as a tax levy for parks.
■ *Capital project funds:* Used to finance capital projects from a variety of revenue sources.
■ *Debt service funds:* To collect funds to repay debt.
■ *Permanent funds:* Used for special purposes on a long-term basis.
■ *Enterprise funds:* Used for separate enterprises to provide services to external parties, such as utility services.
■ *Internal service funds:* Used to account for intergovernmental transactions of services and payments.

The first five of these categories are considered governmental funds, and the last two (enterprise and internal service) are proprietary funds. Other types of funds can handle issues such as special assessments, investment trusts, and other fiduciary purposes.

Accounts for Regulated Utilities

For-profit utilities are "regulated utilities" because they are regulated by state public utility commissions. Many electric companies and water supply companies are in this category. However, fewer wastewater utilities operate as private companies.

The National Association of Regulatory Utility Commissioners (2001) has a uniform system of accounts, and it specifies categories for different classes of utilities. In addition to state public utility commissions, private utilities are regulated by the SEC.

Accounting for Fixed Assets

It is ironic that, on the one hand, infrastructure systems are capital intensive but, on the other hand, accounting for their fixed assets has been neglected in management accounting, which in the past has not produced much information on them. It is one thing to keep financial information on fixed assets on the books, but it is altogether another thing to use this information in management decisions. The situation has been as Peterson (1994) described it in the telephone world: "Put it in, use, if it breaks repair it; if it breaks too many times, discard it and replace it."

Fortunately, this unsatisfactory situation is being corrected—or at least we are on the way, with the introduction of GASB Statement No. 34 (GASB 34), which was adopted in 1999 to identify the costs of acquiring, owning, operating, and maintaining infrastructure. GASB 34 gives governments the choice to adopt traditional methods of calculating depreciation based on historical costs or to adopt an asset-management system.

On balance sheets, assets are classified as current and noncurrent, which include fixed or plant assets that include property, plant, and equipment (Williams 1991). The word "plant" originates from manufacturing accounting. Fixed assets are tangible, have a life longer than one year, and are of significant value. Current assets are more dynamic and have greater effects on tax and profit reports.

No comprehensive document has been published to bring all the concepts of accounting for property, plant, and equipment together (Peterson 1994). Fixed assets are depreciated by accountants, but depreciation relates to tax obligations more than it does to the condition of assets. In fact, in government accounting—including for water, sewer, and stormwater units—the depreciation of fixed assets used to be optional. Now, with GASB 34, accounting for them is required.

Also, with GASB 34, the GASB has created a separate category for infrastructure fixed assets, which are "immovable and of value only to the government unit." The GASB thus requires government entities to report their capital assets on their annual balance sheet and income statement. GASB 34 has set new standards for government accounting for fixed assets; that is, government financial reports must include the costs of asset ownership. This has ended the era when governments assumed that once funds were sunk into physical assets, there was no reason to account for them because the assets would be replaced once they were worn out.

GASB 34 offers two ways to account for the costs of asset ownership. The first method, based on depreciation reporting, estimates the useful life of the asset and deducts a proportional amount each year from the asset value. This is standard depreciation accounting practice, whereby an arbitrary service life is assigned to the asset. This approach does not recognize inflation, and the replacement cost might be higher than the original cost. Also, it does not reflect the condition of assets. Obviously, the accounts will not show accurate obligations to replace and renew the assets.

The second method—the modified approach—tries to correct these deficiencies. In it, the condition of assets is reviewed every three years and rated according to a scale. In this approach, you set goals, measure and report the condition of assets, and report the information, along with all the money spent on maintenance and improvement in annual reports (American Water Works Association 1998).

The implementation of GASB 34 is still in the early stages. It has only been a decade since it was introduced, and no revolution in infrastructure management is apparent. No doubt there has been significant activity among accounting units, but how it will play out in infrastructure investment is not yet apparent.

Thus, fixed asset accounting is required for effective capital management. Peterson (1994) asked the key question: "Do you have in place a process that monitors the current condition, evaluates the future need for replacement, and brings to your attention needs to modify that plan?" He also wrote that "assets must be managed, not just purchased, used up, and replaced." This objective supports the need for asset management (see Chapter 14): to account for assets but also allow management to get the most out of the company's investment. Otherwise, infrastructure suffers from the concept of "put it in place and forget about it."

Financial Statement Analysis

Financial statement analysis can be used to analyze the performance of private sector businesses and government enterprises (Jablonsky and Barsky 2001). The financial statements presented above will be used here to illustrate several useful ratios and indicators that can be derived from them.

Financial Ratios and Market Valuation

The balance sheet shows relationships between the assets under management's control and its responsibility to creditors (through liabilities) and to owners (through shareholders' equity). For public agencies, the same principles apply, but the creditors and shareholders are different. Jablonsky and Barsky (2001) presented a "strategic profit model" with four variables (I have assigned the symbols):

S = net sales
I = net income
A = total assets
E = shareholders' equity

Jablonsky and Barsky presented five ratios of these variables:

I/E = return on shareholders' equity (ROE)
I/S = profit margin
S/A = asset turnover
I/A = return on assets
A/E = financial leverage

Data from the RMIC samples given above are presented in Fig. 10-6 and can be used to calculate these ratios. The 2006 data are from the income statement and balance sheet (Figs. 10-2 and 10-3), and the 1996 data are from the income and cash flow statements for that year (which were not presented here).

The Return on Shareholders' Equity

The return on shareholders' equity is a ratio that shows how a company is performing relative to the equity held by its shareholders. The balance sheet

	1996	2006
Net sales	300,000	500,000
Net income	15,000	84,000
Total assets	720,000	1,100,000
Shareholder's equity	60,000	600,000
Profit margin	0.05	0.17
Asset turnover	0.42	0.45
Return on assets	0.02	0.08
Financial leverage	12.00	1.83

FIGURE 10-6. Sample financial data for Rocky Mountain Infrastructure Consultants

and income statement can be linked to compute ROE as net income divided by shareholders' equity. To do this, we use the accounting equation:

$$\text{Assets} = \text{Liabilities} + \text{Shareholders' Equity}$$

For the case of RMIC, this is for 2006:

$$1,100,000 = 500,000 + 600,000$$

This tells us little, so we have to examine change over the years. We go back to Fig. 10-3, the balance sheet for December 31, 1996, some five years after start-up and 10 years before the current statement. Figure 10-3 shows the comparative data.

The net income is the amount of earned assets that management created during the year. From the income statements,

$$\text{Net income} = \text{Revenue} - \text{Expenses} \ (= \text{net earned assets})$$

RMIC went from a small loss in 1996 to a profit in 2006. We compute the ROE as:

$$\text{ROE} = \text{net income/shareholders' equity}$$

So, for RMIC, the ROE fell from 0.25 to 0.14. The company is not growing much. It grew faster in its first 5 years than in the next 10. Therefore, its ROE has fallen.

Profit Margin
Profit margin is a ratio that shows profitability, and it is calculated as net income divided by sales. It shows the relationship between profit or net income and customer financing through sales.

Asset Turnover
Asset turnover shows how many sales you generate from your assets, and it is calculated by dividing the value of sales by the value of assets. In other words, asset turnover is the ratio between customer financing through sales and the economic resources used to run the business.

Return on Assets
Return on assets is calculated by dividing income by the value of assets. Return on assets shows how much profit you are generating as a ratio of total assets under management control.

Financial Leverage

Financial leverage is calculated by dividing the value of assets by the value of equity. Financial leverage shows the relationship between total assets and the assets provided by shareholders.

Analysis of the RMIC Example

Following the discussion in (Jablonsky and Barsky 2001) we can now interpret these ratios for RMIC, as shown in Fig. 10-7:

- The ROE fell because stockholders' equity was initially low and gave an optimistic picture of income as a percentage of stockholders' equity. As stockholders' equity increased, the ROE fell, even though profit increased.
- The profit margin has gone up. RMIC is becoming more profitable, as a percentage of sales and fees.
- Asset turnover is about the same. Both assets and sales have slowly increased. RMIC has not been able to increase its sales as a percentage of assets.
- The return on assets is up by a factor of 4. Assets have not changed as much as income.
- Financial leverage, the ratio of the total assets to shareholders' equity, fell in the 10 years mostly because shareholders' equity has gone up by a factor of 10.

These ratios would be more meaningful in the context of a large company, as Jablonsky and Barskey (2001) demonstrated using Walmart's data.

	1996	2006
Net sales	300,000	500,000
Net income	15,000	84,000
Total assets	720,000	1,100,000
Shareholder's equity	60,000	600,000
Return on equity	0.25	0.14
Profit margin	0.05	0.17
Asset turnover	0.42	0.45
Return on assets	0.02	0.08
Financial leverage	12.00	1.83

FIGURE 10-7. Financial ratios for Rocky Mountain Infrastructure Consultants

Note: These ratios were calculated from the sample data given in Fig. 10-6.

Market Valuation

Financial ratios are useful in stock valuation, but because RMIC is a closely held company, its stock is not traded publicly. Therefore, no stock price can be established. Let's take a company that is traded publicly—in this case, a fictitious one. Let's say we are studying a stock whose price grew at compound annual growth rate of 10% in the last 10 years. If we want to know the total market value or market capitalization, we would compute it as:

$$\text{Market Cap} = \text{stock price} \times \text{shares outstanding}$$

This actually gives us the distinction used in the stock market between "large cap" stocks and "small cap" stocks. This distinction can change over time, but a large cap stock would be $10 billion or more, and a small cap stock would be about $300 million to $2 billion (Investopedia 2007).

The price/earnings ratio is a useful indicator of a stock's value:

$$P/E = \text{Stock price}/\text{EPS}$$

Where EPS = earnings per share = net income/shares outstanding.

Another convenient ratio is market to book = market value/shareholders' equity. In stock analysis, the book value is the shareholders' equity or the company's assets minus its liabilities.

If a stock's price is low relative to its earnings and book value, it is probably in the category of a "value stock."

Analyzing Government Financial Statements

The financial ratios explained above for the private sector can also be adapted for the context of government finance. Thus, the ratios presented above could be interpreted this way for government finance:

S = net sales = revenues of an enterprise
I = net income = difference between revenues and expenses
A = total assets = sum of current, long-term, and fixed assets
E = shareholders' equity = accumulated value of infrastructure and retained earnings that belongs to ratepayers

The ratios remain the same:

I/E = ROE
I/S = profit margin
S/A = asset turnover
I/A = return on assets
A/E = financial leverage

In this context, we would expect infrastructure systems to have large fixed assets compared with any retained earnings, which are in the form of a reserve fund that can carry a utility through rainy days and earn interest in the meanwhile. Ratepayers and taxpayers would not expect a government enterprise to accumulate more reserves than it needs.

In the case of a government entity, debt is very important, which shows up in shareholders' equity, which is:

$$E = A - \text{Liabilities}$$

So, if debt is the main long-term liability and short-term liabilities are small, one important ratio would be:

$$\text{Debt} / \text{Assessed valuation}$$

and a ratio of 10% might be an appropriate legislative limit. Assessed valuation is one measure of the "customer base." Another ratio might be:

$$\text{Debt} / \text{real market value of tax base}$$

and a ratio of less than 5%, for example, might be an appropriate limit.

These ratios are similar to the financial ratios applied to individual borrowing for home ownership, where debt to income is limited. The ratios vary by locale (Aronson and Schwarz 1996).

Management Accounting and Cost Control

Cost control and management also involve operating and capital items, such as construction costs, operating and maintenance costs, and such other costs as those for regulatory programs and planning. Cost analysis may involve techniques such as value engineering and how to cut waste in the system. Cost analysis is also important when the financing study determines the components of cost that can be assigned to different users.

Costs can be classified as direct or indirect. Direct costs are those directly assignable to the provision of a particular service. Examples are wages, equipment, operation and maintenance expenses, depreciation, and capital expenses. Indirect costs are those that are necessary for the delivery of a service but that cannot be attributed directly to the service itself. Examples would include central services such as computer and support services.

Cost control is a matter of making sure that full value is received for every dollar spent. This is a function of management at all levels, and it requires careful attention to the planning and approval of expenditures, as well as postexpenditure audits to determine how well the investments in program and equipment have paid off.

Regardless of how revenues are developed to build and operate facilities, costs must be allocated fairly to repay and to generate the needed revenues. The allocation of costs requires attention to principles of equity, which always will attract controversy due to the inherent nature of the problem. Cost allocation goes beyond the concept of setting the rate and extends to political questions between levels of government and between the government and private parties about how to allocate and share costs. In recent years, this has become a very important topic as the level of government subsidies has continued to fall for all services.

Cost allocation means to find ways to assess costs in a fair way in proportion to how different parties benefit from a project or a service. This idea was discussed earlier in the book under the topic of user charges (which are covered in more detail in Chapter 12). Most projects and services have costs that are necessary to run the service in general, sometimes called "joint costs," and costs that are clearly identifiable with beneficiaries, sometimes called "separable costs." In the previous sections, those services that lend themselves to "utility management" can often directly focus their costs on the public using the service and levy charges accordingly, but they still must exercise "cost allocation" between classes of customers. Other services are not even able to distinguish who their customers are.

Many examples of cost allocation could be cited. Here, I present three that are often encountered in water management and thus illustrate the general principles involved. For infrastructure services other than water, the same principles often apply; but for most services, costs are easier to allocate than for water.

The first example is the case of the allocation of costs among levels of government to pay for a multipurpose water project. Take, for example, an Army Corps of Engineers multipurpose reservoir located near urban areas in the eastern part of the United States. The project might have, for example, three purposes: water supply, flood control, and hydroelectric power. The federal government has been debating its policies for cost sharing of these purposes for several years, and it is not settled, so the example will be hypothetical in terms of policies. We might say that the hydropower will be produced by the government and sold to utilities on a wholesale basis, so the separable costs of this purpose are entirely financed by user fees. The flood control effort might be jointly financed by the federal and state governments, with appropriate allocation by negotiation. The water supplies might be financed through the sale of the water to local governments using long-term contracts. In this example, then, the allocation of costs is done mostly through the political and negotiation process.

In the second example, the project might be a drainage and flood control project that is necessary to develop a part of an urban area. Some land developers will benefit from the improvements to their property, but some of the benefit will accrue to the public at large. The city will decide on the allocation of costs through negotiation with the developers and with reference to the policies and goals of the city administration.

In the third and final example, the allocation will be according to customer classes—say for a water and wastewater utility. The water cost allocation would be according to standard procedures, and the wastewater allocation would have to consider the impact of industries. A variation of this that has not been used very much is that of the zonal allocation of costs. In other words, if it is more expensive to serve some zones of a city than others, the appropriate rates would be assessed. There are, however, variations in charges between central cities and suburban areas, and this is often the source of distress in water management. When the local considerations of rate needs are added to the local political situation, the result is often complex total rate structures and rules.

It would be nice if cost allocation could be organized so that no negotiations were ever necessary; that would greatly simplify infrastructure management. But that is not likely to happen, because too many actors and profit margins are involved. This is why it is so important for infrastructure managers to be familiar with all the available techniques.

"Triple Bottom Line" Reporting

Infrastructure and environmental enterprises lend themselves to "triple bottom line" (TBL) reporting, which provides a display of achievements and setbacks in the economic, environmental, and social categories. TBL reports can include financial data, but the economic category would address more issues, such as economic development. Environmental and social accounts would address both the positives and negatives for habitat, society, and related issues.

A TBL report can range from a regular comprehensive annual financial report, augmented by social, economic, and environmental results, to a special, focused TBL report that focuses only on the organization's economic, social, and environmental aspects. In that sense, it would be somewhat like what is called a "popular report" in financial reporting.

TBL reporting received its name from the sustainability movement, but infrastructure and environmental managers have also reported economic, social, and environmental impacts in the past. These were often thought of as "planning" reports, whereas the TBL report can be an augmented business report for a utility or agency.

Accounting is the language of business, and a dialect of it is used in public management. This dialect is public sector accounting, with rules about the organization of funds, fixed assets, and public budgets, among others. Accounting takes on different forms for private companies, public companies, regulated utilities, government departments, and private sector nonprofit organizations. Most infrastructure and environmental accounting is in government departments and utilities, which are usually regulated by political boards and by public utility commissions.

Local government agencies prepare their financial records according to rules of the Government Accounting Standards Board, which with its statements provides guides for everything from organizing the books to accounting for infrastructure itself. A utility's or agency's financial statements and annual report tell us a lot about its performance. These include its income statement, balance sheet, and comprehensive annual financial report.

However, because infrastructure and environmental organizations deal with issues beyond monetary profitability, TBL accounting enables them to report social, environmental, and economic results, as well as financials. Though TBL reporting is a developing art, it provides a structure that can enable these organizations to view their total scorecards.

References

American Water Works Association. (1998). *Water utility capital financing: Manual of water supply practices.* American Water Works Association, Denver, CO.

Aronson, J. Richard, and Schwarz, Eli, eds. (1996). *Management policies in local government finance.* 4th ed. International City/County Management Association, Washington, DC.

Block, Stanley B., and Hirt, G. (1997). *Foundations of financial management.* 8th ed. Irwin, Homewood, IL.

Financial Accounting Standards Board (FASB). (2007). Facts about FASB. http://www.fasb.org. Accessed August 26, 2007.

U.S. General Accounting Office (GAO). (1982). *Effective planning and budgeting practices can help arrest the nation's deteriorating public infrastructure.* U.S. Government Printing Office, Washington, DC.

Investopedia. (2007). Large cap. http://www.investopedia.com/terms/l/large-cap.asp. Accessed July 18, 2007.

Jablonsky, S., and Barsky, N. (2001). *The manager's guide to financial statement analysis.* 2nd ed. John Wiley & Sons, New York.

Miller, J. R., and Warren, S. (1991). "State and local government accounting." In *Accountant's handbook*, 7th edition, ed. D. R. Carmichael, Steven B. Lilien, and Martin Mellman. John Wiley & Sons, New York.

National Association of Regulatory Utility Commissioners. (2001). http://www.naruc.org. Accessed June 24, 2001.

Peterson, R. H. (1994). *Accounting for fixed assets.* John Wiley & Sons, New York.

Reilly, D. (2006). "Accounting cop takes the beat at pivotal time." *Wall Street Journal*, June 24.

Williams, J. R. (1991). "Financial statements: Form and content." In *Accountant's handbook*, 7th edition, ed. D. R. Carmichael, Steven B. Lilien, and Martin Mellman. John Wiley & Sons, New York.

11

Public Finance and Budgeting

Public and Private Sector Accounting

Public finance is the field that controls the management of public funds for infrastructure and regulating the environment. Public sector budgeting is a critical financial tool and adds transparency to government functions.

Infrastructure and environmental problems require the use of public finance, which entails important differences from private sector finance. Whereas in private business, making a profit is an important goal, in the public sector the goal is to provide a high-quality mandated service in a cost-effective way. To respond to this, public finance has become a recognized field with its own standards, associations, and reference materials.

In public finance, government activities that serve public purposes are controlled through budget processes with legally mandated spending boundaries. The fund accounting process provides accountability that the government entity is complying with the budget plans. Private businesses do not require fund accounting, and this is a significant difference in the two accounting systems.

This chapter provides an overview of public finance and the all-important public agency budget process. It illustrates how infrastructure and environmental managers can learn public finance and use it and also learn the fine points of budgeting for local, state, and federal government work. It also introduces the reader to financial tools that can be used to overcome barriers to good government, a topic explored in more detail in Chapter 17.

Federal, State, and Local Finances

Public finance is concerned with actions by government at all three levels. Federal financial management attracts the most headlines, particularly for big ticket items like war and Social Security. However, state and local government financial issues are increasing in importance, especially for infrastructure and environmental managers.

As a policy instrument, the federal budget is the channel through which many national policy decisions flow and thus has global impacts. Global macroeconomic effects occur through instruments such as tax policy, government expenditures, debt and deficit management, national funds, and congressional appropriations.

State and local budgets also have large impacts, but they are distributed more thinly across many locales. State government financial issues are becoming more important as states increase in size and sophistication. California already has one of the largest economies in the world, comparable to countries about the size of Spain. Other large states like New York and Texas also have large and significant budgets. Budget politics in state governments include tax-limitation initiatives such as the 1978 Proposition 13 in California and Colorado's 1992 Taxpayer Bill of Rights. Financing state government is a continuing issue that includes contentious questions such as the use of lotteries. Local finance is closer to home, relying mostly on property and sales taxes and on dedicated fees. Local government finance includes revenues for utility services, which are usually larger than general taxes.

Financial controls in the federal government involve the Department of the Treasury, the Internal Revenue Service, the Office of Management and Budget (OMB), the Government Accountability Office (GAO; formerly the General Accounting Office), the Congressional Budget Office, and other entities.

State and local governments have smaller units that perform the same financial functions. For example, the OMB controls budget functions for the federal government, but state governments also have budget agencies, and local governments have designated budget functions within the mayor's or city manager's office.

Regulatory Control of Public Finance

Public finance is controlled by a different set of regulatory authorities than private sector finance, which is governed by boards and commissions such as the Federal Accounting Standards Advisory Board and the Securities and

Exchange Commission. The principal regulatory offices and programs for government finance are summarized in this list:

- The Congressional Budget Office performs analysis of budget for legislative branch.
- The Federal Accounting Standards Advisory Board sets accounting standards for federal agencies.
- The GAO is part of the Office of the Comptroller General of the United States.
- The Government Accounting Standards Board sets accounting standards for state and local government organizations.
- The Government Performance Results Act of 1993 introduced new controls on and reporting requirements for how government funds can be spent.
- The OMB controls the federal budget. State and local governments have budget offices with comparable roles.
- The U.S. Department of the Treasury handles borrowing for the federal government.

Government Revenues: Tax Policy, Fees, and Tax-Limit Initiatives

Government revenues come from taxes, fees, borrowing, and intergovernmental transfers, whereas private sector revenues come from the sale of goods and services. Both government and private organizations can have investment earnings. In nonprofit private enterprises, revenues also come from donations and other diverse sources. The approximate distribution of revenues among the three levels of government is as shown in Table 11-1.

Tax policy addresses the ways that the government decides how much of the public's money it gets and for what reasons, including what types of taxes are best from the standpoint of equity and efficiency. The main academic fields that study tax policy include economics, political science, and law. Tax policy is one of the most powerful tools of government policy and is a constant source of conflict. There are many kinds of taxes at all three levels of government, so it is hard to generalize. However, we can briefly discuss tax policy at the federal, state, and local levels of government.

The Federal Level

Chapter 2 reported that federal tax receipts for 2006 were $2,416 billion, distributed as individual income taxes, $1,096 billion; corporate income taxes, $261 billion; social insurance and retirement receipts, $884 billion; excise taxes, $75 billion; estate and gift taxes, $24 billion; customs duties,

TABLE 11-1.   The distribution of revenues from taxes and charges across government levels, 2006

Government level	Tax and charge revenues (billions of $)	Percentage of total
Federal	2,400	65
State	700	19
Local	600	16
Total	3,700	100

Sources: U.S. Bureau of the Census 2007a, 2007b.

$28 billion; and miscellaneous receipts, $48 billion. This indicates that the government has decided that most of its activities should be financed by individual wage earners and business corporations, and the other categories of taxes are aimed at special purposes, such as Social Security and import duties. From the basic economic flows that were shown in Fig. 2-1, you can see that the government does not have much choice in this basic decision, but it has more choices in the details of these tax schedules, especially for individual income taxes (OMB 2006).

An income tax places a levy on the incomes of persons and organizations. Tax incidence, or how a tax affects different groups, can vary according to policy. The tax schedule can proportionately require more revenue from high earners (i.e., a progressive tax), distribute the tax burden at fixed rates regardless of income (a proportional tax), or place more of the burden on low- and middle-income earners (a regressive tax). A capital gains tax is also a type of income tax.

The United States did not have an income tax until 1861, when Congress passed a law mandated a 3% tax on all annual incomes above $600. The tax was on again and off again until the 16th Amendment was ratified in 1913. Today, the rates for individuals are between about 10% and 35% of income (U.S. Department of the Treasury 2008). Income tax policy continues to be a source of much debate in the United States. Some advocate a flat tax to get rid of all the deductions, exceptions, onerous record keeping, and other negative aspects of the current tax system. Another policy issue is whether tax cuts can stimulate the economy and put more money into the hands of wage earners.

The State Level

State tax collections in 2006 were $706 billion. This is approximately 30% of the total federal collections, or 46% of federal collections other than Social Security. Property taxes accounted for about $12 billion; income taxes, $292 billion; sales taxes, $330 billion; licenses, $45 billion; and other collections, $28 billion (U.S. Bureau of the Census 2007a).

Strong efforts have been made to limit state and local tax collections. For example, in Colorado the 1990s Taxpayer Bill of Rights placed constitutional limits on state and local tax receipts. This has placed very tight restrictions on funding for infrastructure, as well as other government programs, including education. Colorado voters passed a "time out" measure in 2006 to get some relief from this measure.

The Local Level

For 2006, state and local tax totals were $1,205 billion, making the local share some $500 billion, comprising mostly sales and property taxes. These numbers vary somewhat with the source. The 2002 Census of Governments reported $597 billion in local government general revenues, but this included charges and fees. Utility revenue added almost another $100 billion, including water, electric, gas, and transit services (U.S. Bureau of the Census 2007b).

Government Borrowing

Chapter 2 explained how persistent budget deficits have created a national debt of more than $9 trillion, or about 70% of gross domestic product. This debt is guaranteed by government obligations such as Treasury bonds and Treasury bills, as explained in Chapter 13. With prevailing interest rates in the range of 5%, you can see that the interest payments on this debt are in the range of $450 billion, if all had to be paid at the 5% rate.

Although this is a serious matter, inflation erodes this debt over time. Imagine that the value of this debt represents bond investments by individuals. Each year, they get 5% in interest payments, but the value of their holdings decreases 5% due to inflation. If future budgets are balanced, then the government's obligations will be declining relative to its ability to pay by taxing the gross domestic product. This simple way of looking at this situation illustrates that inflation gives the government a tool for fiscal policy through borrowing.

Government Enterprises

Government enterprises are sometimes used to provide services for which charges can be imposed, like utility services. These can be authorities, utilities, departments, and other units. At the U.S. national level, the Tennessee

Valley Authority is an example of a government enterprise. A state-level government enterprise might be a toll road authority. At the local level, a city water department might be a government enterprise.

In developing countries, these enterprises are sometimes called state companies or state enterprises. For example, in Mexico the state oil company is called Petróleos Mexicanos (PEMEX). In Brazil, the national electricity company is Centrais Elétricas Brasileiras (Eletrobrás).

The Enterprise Principle

Whatever the infrastructure organization, if it can be supported by its own revenues, it follows the "enterprise principle" and thus does not depend on subsidies from general purpose taxes. The concept is that services should be self-supporting and charged according to the benefits users receive from them. The practice of pricing through user charges is the basis for the control of the allocation of the services and for raising revenue. The equity issue is central to the philosophy, that the charging schemes should be fair.

If a service is self-supporting, its revenue generation activity and financial control are under the supervision of its manager rather than the political process. However, the desires of customers must be factored into the manager's decisionmaking.

Subsidies cannot always be avoided, as in the case of providing vitally needed services that cannot pay for themselves. The use of subsidies for pubic transit is common, for example, because the fare box does not pay the full bill. The federal government has been providing operating subsidies for transit systems on a general basis for a number of years. Other examples of subsidies are the construction grants program for wastewater and the construction and operation of public housing.

The use of subsidies in developing countries is widespread, often providing the difference between life and death. In the case of irrigation systems in developing countries, for example, even though operation is not directly subsidized, it is indirectly supported through the allowance of deferred maintenance, with "catch-up" grants and soft loans for rehabilitation.

Budget Processes

Budgeting as a Management Tool

In public finance, budgets are more than just a way to track expenditures; they are powerful management tools for organizations. Actually, public budgeting has not been around that long. At the federal level, the OMB was created as the Bureau of the Budget by the Budget and Accounting Act of 1921.

Originally located in the Treasury Department, the Bureau of the Budget was moved to the Executive Office of the President in 1939 and renamed the Office of Management and Budget in 1970. The New York Bureau of Municipal Research is considered a pioneer in municipal budgeting for its work in the period 1907–15 to provide reforms through more open budget processes and controls (Moak and Hillhouse 1975). Budgeting at all three levels of government has continued to evolve. The last 25 years have seen a large change in intergovernmental relations, and local governments are now much less dependent on the federal government (Mullins and Pagano 2005).

A budget is an adopted plan for expenditures and revenues structured to follow the programs and divisions of an organization. It is a policy document, operations guide, financial plan, and communications medium. Budgeting involves decisions about an organization's policies and directions. One decision is the level of charges or taxation. In local governments, this is the portion of community resources needed for governmental programs and services, or the "cost of government." Budgeting shows the emphasis that is to be placed on different programs within the governmental structure. Within programs, the budget shows how money will be allocated to personnel, equipment, contracts, and other expenditures. The budget also specifies sources of revenue, whether from debt, user charges, or other sources.

At the federal level, the budget is overseen by the OMB, a large agency with many fiscal analysts. In a city government, a budget office in the city finance office will perform the same functions. In a city, the budget officer may report to the finance director and be responsible for the city's operating budget, capital improvement program, and financial plans.

The Budget Process

Preparing the operating and capital budgets for a government unit is known as the "budget process." It comprises planning, negotiating, presenting, adopting, following, and auditing the budget for the organization or program. Budgets link with planning through a planning-programming-budgeting system, which goes on continually. The plans and programs become reality when they are translated into budget actions.

The budget is planned in advance. As the budget year approaches, budget negotiations become detailed and nail down decisions about what the organization will do. After the budget is authorized, it becomes the official plan for the operation of the program for the designated fiscal year. Before the budget is adopted, it is the "proposed budget." During the budget year, the budget is used to control expenditures and regulate the organization's activities. Budget information is used to control expenditures so they stay within budget. After the year closes, the budget can be used to do a post facto evaluation of performance.

Both the operating and the capital budgets are planned on multiyear cycles. The capital budget is linked to comprehensive planning and needs assessment processes. The operating budget is linked to plans for services, organizational development, and programs. In the year that the budget is spent, the funds used are those approved during the previous fiscal year.

In the budget planning year, the organization submits estimates of the funds that will be needed for a multiyear period. In the budget preparation year, detailed planning leads to approval of the next year's budget by the governing board. In any year, there are at least three budget years in the manager's life: one for planning, another for approval, and the third for operating.

The budget calendar determines the sequence of activities for the budget and disciplines the process. Figure 11-1 shows the general budget cycles for the federal and local levels of government.

Operating and Capital Budgets

Most organizations have one budget for capital and another for operations. The exception is the U.S. government, which has a unified budget that does not distinguish between capital and operating expenses.

In the operating budget, the details of ongoing expenses and revenues are projected, approved, and reported. Examples would be personnel costs, fuel, rent, and other recurring expenses. The operating budget provides information to aid in planning and requires administrators to produce an estimate of expenditures so that the adequacy of revenues can be checked. It also provides a means for managers to evaluate the internal competition for resources, and it is used in work planning and evaluation. It is used to communicate operating objectives to the policy oversight body and to make adjustments required by them. In addition, it provides a basis for conducting a financial audit of the enterprise.

The budget document normally contains a message of the chief executive, an estimate of projected revenue, a summary of proposed expenditures,

	Pre-planning year	Budget approval year	Budget year	Budget audit phases
Federal cycle	Agency plans OMB analysis	President's message Congressional action	Agency spending Supplementals	Financial and performance audits
Local cycle	Preliminary department and program plans	Chief executive message Council actions	Department spending	Books close Audits

FIGURE 11-1. The general budget cycles

Note: OMB is the Office of Management and Budget.

comparisons of finances in years past, and other parameters of interest to management, policymakers, and customers. The resolution by the policy group to approve the operating budget provides the chief executive with the authority to spend for the organization.

In a sense, the capital budget is a subset of the operating budget. Capital budgeting considers longer time spans for capital items such as facilities and equipment. Construction projects, equipment, and the acquisition of real property are financed through capital budgets. The GAO has prepared several reports on capital budgeting practices. It has defined capital budgeting as "the way organizations decide to buy, construct, renovate, maintain, control, and dispose of capital assets" (GAO 1981; also see GAO 1982).

A five-year capital program would show capital expenditures for five years in the future, beginning with the next budget year. The items in the first of those years should be the same as the capital requests in the current capital budget.

Figure 11-2 shows the relationship between the comprehensive plan, the capital investment program, and the capital budget.

Federal Capital Budgeting

Although some think capital budgeting is a good idea for the federal government, others think it would only lead to more pork barrel politics. Studies have recommended a federal capital budget, but it has not happened. For example, one of the early infrastructure policy studies contained this recommendation: "Congress should mandate the creation of a coordinated national infrastructure needs assessment program and, within the unified budget, require that capital expenditures be presented and highlighted in a clear, comprehensive way" (University of Colorado 1984). This recommendation was justified in this way: "No easy, clear way now exists to measure the full extent of federal commitment to infrastructure investment. The federal government, through a variety of investment strategies, provides considerable support for infrastructure development. But it is impossible

FIGURE 11-2. The comprehensive plan, capital investment program, and capital budget

to determine which approaches are most effective. The Advisory Committee urges Congress to assure that capital expenditures are separated from current operational outlays within the unified federal budget. Congress can then debate and set capital priorities separately and deliberately."

Recommendations like this go back at least to the Hoover Commission in 1949 and have support from economists and fiscal analysts (Hofman and Cook 1982). The arguments against a federal capital budget are that special interests will have a heyday with it; that the information needed from a capital budget is already provided by "special analysis D," a section of the budget documentation; and that everyone would want their program classified as an "investment" in the capital budget rather than placed in the operating budget (Boskin and Ballantine 1986).

Government Budget Processes

Pubic budget processes differ between government levels. Naturally, the federal process is more complex than those at state and local levels. At the national level, there are many more interest groups and types of programs as well as budget categories. Decisions about the federal budget have far-reaching impacts on economic health and even on international matters such as the strengths of nations' currencies. The federal government has been in deficit spending for a number of years, and debt is an instrument of economic policy.

The cycle of the federal budget year begins on October 1. The year of budget planning is intense, as agencies compete for the right to submit budget requests. The OMB reviews the requests and may block them from going to the president. The process culminates with the president's annual budget message to Congress in early February. This is the beginning of congressional consideration of the budget, which is supposed to conclude by the beginning of the federal fiscal year on October 1. In practice, the process may not be completed by then.

State and local governments have different budget cycles, but the basic processes are the same. The agencies prepare provisional budgets, which are reviewed by a budget office. Recommendations go the chief executive, who submits the budget request to the approving body. In state governments, approval is normally given by the legislature. In local governments, it is given by the city council or board of directors.

The Politics of Budgeting

Public sector budgeting involves much political maneuvering because so much money is involved. Competition can be internal or external (Wildavsky 1984). External lobbying for the budget occurs in keeping with the "iron triangle" phenomenon explained in chapter 1. Interest groups seek budget appropriations to benefit from programs or increase their influence. They may even have paid lobbyists working to increase their budgets. Client groups at the federal level range from the elderly, who have a great interest in Social Security, to environmentalists, who will lobby for more money to build wastewater plants. At the local level, developers watch the budget process to determine how much they will be expected to pay in infrastructure fees.

The internal politics of the budget process have to do with gaining power and influence. Employees and managers may seek status and power within a public organization by growing their budgets.

Deciding how much the unit will request is one tactic in budget politics. At lower levels, managers may ask for what they need and perhaps more, knowing what they propose will be cut. Deciding how much to spend is the role of the agency's executive leadership, which must pare down competing requests. In infrastructure organizations, it is always easy at budget time to defer capital items that have long-term implications but few short-term consequences.

The budget office has a tough job in dealing with line managers at all levels of government. Budget officers must support the goals of top management, and these may not be the same as those of line managers. The budget office will decide how much to recommend to the approving authority. This will be different in a local situation than in the complex world of federal government politics.

At the state and federal levels, the perspectives of appropriations committees will be important. This is not a factor at the local level, unless the governing board must deal with a budget committee.

The policy organization decides on appropriations. It faces constraints in the form of available revenues and debt levels. The federal government is the only government level that is normally allowed to go into debt, with the exception of government enterprises like utilities.

Budget methods continue to evolve, and a new approach is called "budgeting for outcomes." This is actually more than just a budget process and extends to a change in the philosophy of how government is run. Chapter 17 explains budgeting for outcomes in more detail.

Capital Improvement Programming and Budgeting

The planning and implementation of civil infrastructure systems require valid plans that flow from the most general, integrated plans to more specific capital plans for categories and sectors of infrastructure systems. Planning for facilities, which is called "capital improvement planning," includes the installation and renewal of the infrastructure required for growth and ongoing operations. Capital improvement planning requires a series of steps or stages:

- Integrated planning. This requires multidisciplinary work to include population, economic, environmental, political, and scientific assessments and analyses.
- Dividing up responsibilities for sector plans; a sector can be an area (e.g., part of a city) or a function (e.g., transportation or water). This requires organizational work to determine roles and responsibilities, sometimes in an intergovernmental framework.
- Sector planning for areas or functions.
- Deriving the broad outlines of required capital improvements (e.g., a road extension or widening; new roads; rail; a new airport; new water or wastewater facilities; renewals and expansions). This involves scheduling and planning the process of the capital improvement program (CIP)—including the strategy for the CIP, what it will include, who will be involved, how it will be presented, and the like.
- Isolating a set of projects or systems for further planning. Here target systems will be isolated, and the selection will depend on things like urgency, political support, and ability to finance.
- Dividing the projects into subprojects or incremental project stages. This involves engineering to identify logical ways to assemble the CIP packages. For example, a bridge can be built; but if there is no road, it languishes.
- Preliminary planning for projects, leading to costs and other detailed information. This involves engineering work to rough out designs, get cost information, review and coordinate plans, and so on. This step is detailed and costly, so it ought not be undertaken until appropriate.
- The programming of subprojects for years of construction and implementation. Engineers can assess how long it will take to design and build systems, how long it will take to gain approval, when systems are needed, and so on.
- Gaining approval for these elements of the CIP. Approval is often much more difficult than engineers think it will be, and it involves the public and decisionmakers, sometimes requiring elections. Planners must keep in mind the importance of the approval process.
- Determining methods to finance the capital budget. Unless projects can be financed, they may ultimately be useless. Methods such as bonds, loans, direct user charges, and sale to private owners should be assessed.
- Publication of the CIP and inclusion in the entity's capital programs and budgets. At this stage, responsibility shifts mostly to the budget and financial staff and to executive officers.

Applications to Infrastructure and the Environment

The infrastructure or environmental manager will find many uses for public finance, even while not realizing that this is what he or she is learning on the job. This can begin early on with the manager's involvement in the budget process, even if it is something basic like compiling a list of needs. This can then lead to deeper knowledge of why fund accounting works as it does and then move on to the rules for managing funds and so on.

If the manager is in a freestanding government department or separate utility, it might operate under the enterprise principle and require the manager to exercise more fiscal accountability than he or she has experienced before. This can start with the capital program and then lead to detailed work with the operating and capital budgets.

The manager will recognize sooner or later that the budget process involves a lot of politics, and he or she might be glad to know that others have experienced the same thing. This can lead to deeper understanding of the politics of budgeting and to being ready for innovative methods, such as budgeting for outcomes.

At higher levels, the manager will began to understand the nuances between levels of budgeting, to include local, state, and federal governments. Political moves such as taxpayer revolts will come to light, and the manager will begin to see possibilities in the realm of public finance that will enable him or her to overcome so many barriers to good government.

References

Boskin, M. J., and Ballantine, G. (1986). "Does Washington need a new set of books?" *Wall Street Journal*, December 2.

U.S. General Accounting Office (GAO). (1981). *Federal capital budgeting: A collection of haphazard practices*. U.S. Government Printing Office, Washington, DC.

———. (1982). *Effective planning and budgeting practices can help arrest the nation's deteriorating public infrastructure*. U.S. Government Printing Office, Washington, DC.

Hofman, S., and Cook, M. (1982). "Crumbling America: Put it in the Budget." *Wall Street Journal*, October 7.

Moak, L. L., and Hillhouse, A. M. (1975). *Concepts and practices in local government finance*. Municipal Finance Officers Association, Chicago.

Mullins, D., and Pagano, M. (2005). "Local budgeting and finance: 25 years of developments." *Public Budgeting & Finance*, 25(4s), 3–45.

U.S. Office of Management and Budget (OMB). (2006). Budget of the United States government. http://www.whitehouse.gov/omb/budget/fy2006/.

University of Colorado. (1984). *Hard choices: A report on the increasing gap between America's infrastructure needs and our ability to pay for them, report for the Joint Economic Committee of Congress.* University of Colorado, Boulder.

U.S. Bureau of the Census. (2007a). State government tax collections. http://www.census.gov/govs/statetax/0600usstax.html. Accessed July 19, 2007.

———. (2007b). 2002 Census of governments: Table 2—Local government finances by type of government and state, 2001–02. http://www.census.gov/govs/estimate/0200ussl_2.html. Accessed July 19, 2007.

U.S. Department of the Treasury. (2008). History of the U.S. tax system. http://www.treasury.gov/education/fact-sheets/taxes/ustax.shtml. Accessed May 23, 2008.

Wildavsky, A. (1984). *The politics of the budgetary process.* 4th ed. Little, Brown, Boston.

12

Revenue Sources for Infrastructure and Utilities

Revenue to Build and Operate Infrastructure and Utilities

Whereas there is only one federal government and 50 state governments, some 87,000 local governments must finance their operations, many of which deal with infrastructure and utility services. These include municipal governments and a variety of special districts managing infrastructure and utilities for transportation, water supply, and diverse other services. In addition, a large number of private sector utilities manage infrastructure, and they face similar challenges to local governments. Governmental and for-profit utilities use similar service fees. The difference is that certain tax and fee revenues are restricted to governmental organizations.

The federal government is also involved with infrastructure finance, and it operates programs such as the Highway Trust Fund, revolving loan assistance, and energy enterprises. The state governments have departments of transportation, which collectively manage hundreds of billions of dollars every year.

This chapter explains how the operating and capital costs of infrastructure and public utility services are funded through combinations of revenue streams. The focus is on local infrastructure and utilities. National investments in infrastructure were discussed in Chapter 5.

A General Revenue Model

The main sources of revenues for the operations and capital expenses of infrastructure enterprises and agencies are user fees and public taxes. If an agency is supported by a general fund, the source of revenues is taxes. If it is an enterprise, it should be self-supporting from fees and charges. Capital revenues sometimes involve debt financing through borrowing, but borrowed funds are repaid from fees and taxes.

With the possible exception of local transportation, most of the infrastructure and public services discussed in this chapter can be financed through self-supporting enterprises. Their organizational forms range from small utilities to giant authorities, such as the Panama Canal Authority. Regardless of their size, their general finances can be explained by the field of utility finance, which has a well-developed body of knowledge. The reason for the exception in the case of transportation is its heavy reliance on taxes.

Figure 12-1 illustrates typical revenue categories for a water and wastewater utility. Note the division of the system expansion (capital) and operations sides of the organization. The main source of funds for expansion is the developer or builder, and the main source for operations is the customer. However, the customer ultimately pays for expansion as well as ongoing operations.

On the expansion side, the developer or builder finances the system's expansion through plant investment fees (PIFs) and cash in lieu of water rights. PIFs, which are impact or growth fees, pay for the infrastructure, and cash in lieu of water rights pays for acquisition of new water rights for the growing population. This requirement to pay for water rights is a unique feature of some western water systems because the right to use water is a property right.

On the operations side, customer fees in the form of water rates pay for operation and maintenance, as well as some administrative costs. Note that facility replacement is included in the operations charges. This capital charge could as well have been placed with expansion charges to group the capital charges together. Its placement illustrates one of the policy questions in utility finance: To what extent should the renewal of infrastructure be considered within operational categories and/or capital categories?

Note also the flow of funds from the fees toward debt service. This illustrates how the customer actually pays for capital charges in the end. Even when a developer pays a PIF capital charge, the customer repays that by paying for the PIF in the purchase price of the home.

Because the utility is a self-supporting enterprise, it receives no tax subsidies. It actually pays to the general fund in the form of a PILOT (payment

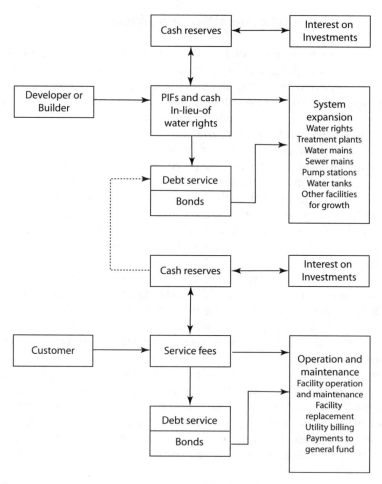

FIGURE 12-1. Utility financial flows

Note: PIFs are plant investment fees. This diagram was originally drafted by Michael B. Smith, director of the water and wastewater utility of Fort Collins, to explain to the Water Board and City Council how the utility's finances worked.

Source: City of Fort Collins 2007.

in lieu of taxes) and an administrative charge that compensates general government for services rendered, such as supervision and accounting.

Some public works services are normally financed by general taxes and not by utility fees. So, within a local government, you may have a mixture of tax-supported services and utility services that are financed by user charges.

Revenue by Level of Government

All three levels of government receive revenues for infrastructure and public service programs, but most of the activity for infrastructure and utilities is at the local level. At the federal level, infrastructure and environment programs involve agencies such as the Federal Highway Administration, the Army Corps of Engineers, and the Environmental Protection Agency. Though they receive diverse sources of funding, most of their work is supported by direct federal appropriations that are derived from taxes. Other funding sources include cooperative cost-sharing programs, such as the Corps of Engineers' cost shares for construction and the U.S. Geological Survey's cooperative stream-gauging programs.

Federal appropriations are supported by tax receipts, and most revenue is from the income tax. In 2007, these receipts were estimated at $2.5 trillion, with $1.17 trillion from individual taxes and $342 billion from corporate taxes. The additional categories of revenue include Social Security ($873 billion), excise taxes, the gasoline tax, various tariffs, and many others (U.S. Office of Management and Budget 2007).

At the state government level, revenue sources include individual and corporate income taxes, property taxes, gasoline taxes, and miscellaneous others. Table 12-1 presents examples of taxes collected in three states (U.S. Bureau of the Census 2007).

Local government finance is more difficult to track than federal and state finances because there are so many local governments (some 87,000, including special districts, compared with 50 state governments and one federal government). The main sources of local government revenue are property taxes, sales taxes, and charges. As an example of local revenue, my

TABLE 12-1. Tax revenue data for selected states

Type of tax or measure	Colorado	North Carolina	New York
Population, million	4.7	8.7	19.0
Total revenue, $ billion	7.6	18.6	41.7
State tax per capita, $	1,620	2,140	2,190
Income tax, $ billion	4.1	9.7	25.9
% individual	92	87	90
Sales tax, $ billion	3.1	7.6	13.3

Note: The sales tax figures include motor fuel. The remaining taxes include motor vehicles, licenses, and miscellaneous other taxes.

Source: U.S. Bureau of the Census 2007.

city of Fort Collins, Colorado, collects the taxes and fees (based on its 2005 adopted budget) shown in Table12-2. The major share of revenue is from fees and charges, mainly in the utility sectors. Sales and use taxes make up 81% of the total tax. In the case of the general fund, the property tax contributes 15%. Most of the property tax goes to the school districts and county government.

Tax Revenue

The two main taxes that fund infrastructure at the local level are the property tax and the sales tax. The gasoline tax, which flows to the federal and state governments, is a type of sales tax. The income tax is the main federal revenue source, and part of it is returned to state and local levels for use in infrastructure projects and for environmental protection.

Property Taxes

Property taxes are a way to tax economic assets and obtain public revenues from those who have wealth. They are mainly levied against residential, commercial, industrial, agricultural, and vacant land and properties, as well as on natural resources, mines, and oil and gas deposits. They are mostly used by local governments, but state governments sometimes share in them.

The property tax is called an "ad valorem" tax because it is calculated according to the value of the property. States use different formulas to apply a "mill levy" against the assessed valuation of property. A mill is 1/1,000 of the assessed valuation, which is the valuation of the property against which the tax is levied.

TABLE 12-2. Taxes and fees for the City of Fort Collins

Budgeted income	Amount (millions of $)
Fees and charges	222.4
Sales and use tax	90.5
Intergovernmental transfers	16.0
Investment earnings	5.6
Contributed capital	10.7
Miscellaneous	8.9
Transfers from other city funds	93.6
Total city revenues	447.6

Source: City of Fort Collins 2007.

For example, a special district with a 1 mill levy on 10,000 properties with an average assessed value of $100,000 each would collect:

$$\text{Revenue} = 1/1000 \times 100,000 \times 10,000 = \$1.0 \text{ million,}$$
$$\text{or } \$100 \text{ per property per year}$$

States may have formulas to adjust the assessed valuation to a level less than the market value. In Colorado, for example, state law currently requires that taxable value be calculated from market value by applying an "assessment rate." The rate is set by the Legislature and was 0.0796 in tax year 2004. If, for example, a $200,000 home was located where the mill levy was 75.541, the annual tax would be (Colorado Division of Property Taxation 2004):

$$\text{Tax} = 200,000 \times 0796 \times .07541 = \$1,202.61$$

Property taxes are usually collected by county governments and distributed to local governments with taxing authority. For example, in 2002, the property tax in Fort Collins of 87.1 mills was distributed this way: 9.8 mills to the city, 22.4 mills to the county, 51.7 mills to the school district, 2.2 mills to the health services district, and 1.0 mill to the water district (City of Fort Collins 2007). In that same year, property within the city had the distributions shows in Table12-3. From this table, you see that residential and commercial property pay the bulk of the taxes.

Sales and Use Taxes

Sales taxes are a way to tax economic activity and obtain revenues from those who have current income, and then to redistribute it to public purposes.

TABLE 12-3. The assessed valuation of property in Fort Collins

Type of property	Actual value (dollars)	Assessed valuation (dollars)
Vacant	166,042,900	48,160,100
Residential	7,104,122,250	650,946,220
Commercial	1,310,901,170	380,184,720
Industrial	61,440,449	176,178,330
Agricultural	5,076,920	881,130
Natural resources	35,380	10,360
Oil and gas	18,208	5,280
State assessment	116,458,667	33,773,150
Totals	9,317,059,875	1,292,139,290

Source: City of Fort Collins 2007.

Excise taxes are a form of sales tax, usually levied against specific items with designated percentages. A use tax is also like a sales tax, but it might be levied on an item bought somewhere else and not subjected to sales tax or on an item that is leased, like an automobile.

To demonstrate the impact of changes in sales and use taxes in Fort Collins, Fig. 12-2 shows the growth and flattening of taxable sales and sales tax collections in the city. As you can see, the city was on a growth trend until about 2001, when a number of negative factors hit, particularly the construction of competing retail space outside the city limits and thus out of reach of the city's tax collectors. By the time the 2005 data were in, the city's total loss in revenue was around $10–15 million, which had a significant impact on its budget. This was the main reason that the city adopted its "budgeting for outcomes" plan, which is explained in Chapter 17. There was a need to cut jobs and services to compensate for the lost sales tax revenue.

Tax Increment Financing

Tax increment financing is a way to use a development district to capture that part of property tax revenues that is due to growth. When a development district or agency implements a plan, it may cause an increase in property values and thus in the tax revenues generated by the property. This tax increment can

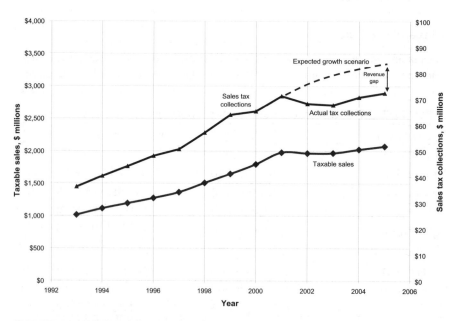

FIGURE 12-2. Taxable sales and sales tax collections in Fort Collins

be reinvested in redevelopment or used for other purposes. For example, in Denver-area redevelopment, the urban renewal authorities are emphasizing "new urbanism" and "transit-oriented" development. Tax increment financing is being used to subsidize some of this new development. Proponents of this trend point to the advantages of more compact walkable communities and getting away from urban sprawl. Opponents think that tax subsidies not voted on by citizens are a bad idea (Lang 2007).

Rates and User Charges

Charges for Public Services

Rates, charges, and fees—such as by electric, gas, and water utilities—are a primary source of user financing for services and infrastructure. The basic theory is that rates and user charges are set to recover the "cost of service" for a public utility. The process is to determine the needed level of service, how much it costs, and how to allocate the costs across customer classes, using some type of equitable scheme. If the utility is regulated by a public utility commission, then in theory these decisions are reviewed and approved by objective third parties.

Although these rational economic principles for setting rates make sense, rate setting also involves complex combinations of law, politics, equity, and business strategy. For one thing, many rates are not regulated by public utility commissions, and even when they are, how those involved determine equity and the cost of service involves complex calculations. Rate setting extends into other areas as well, including congestion pricing and environmental management, where charges can be used to ration environmental resources.

Theories of Rate Setting

The theory of rate setting considers variables such as the type of public good or service and why it is needed, whether it is optional or essential, the extent to which it benefits the general public or only individuals, whether its use can be measured and rationed, and whether it is a natural monopoly. For example, electric power and water are different types of essential services. Other services, such as cable TV, are optional.

Some services have direct benefits for individuals, and others serve distributed public purposes and may merit tax subsidies to spread charges across taxpayers. A service may be individual, such as basic water service, but be required for public purposes such as for human rights and community sanitation. Sewerage service serves individuals and benefits communities.

Residential telephone service benefits individuals but also knits communities together in functions such as emergency response.

Although rate setting differs across sectors of infrastructure and public services, a common set of principles for rates can be presented (Vaughn 1983):

■ *The user pays:* Fees should be levied on the beneficiaries of the services.
■ *Efficiency and equity:* Fees should provide both economic efficiency and equity (efficiency means no waste, that the public gets what it pays for, and that the use of service is rationed; equity assures justice and fairness in access to and the cost of a service).
■ *Marginal-cost pricing:* Prices or fees should be set at the marginal or incremental cost of providing the service, not the average cost (this is an efficiency principle and must be balanced with equity considerations).
■ *Peak-load pricing:* Peak-load pricing should be used to manage demand.
■ *Access:* Access to services should be provided for low-income residents where burdens will result from marginal-cost pricing.
■ *Responsiveness:* User fees should be responsive to inflation and to economic growth.

Arguments Against User Fees

Some groups oppose user fees as instruments to charge for public services. Their arguments focus on the obligation of government to provide services more uniformly and not only on the basis of a charge. These arguments include (Vaughn 1983):

■ *Social benefits:* Services bring social benefits that cannot be measured and charged for.
■ *Income distribution:* Tax payments for services redistribute income to those who cannot afford vital services.
■ *Economic development:* Public facilities and services attract economic development and tax revenues to help finance services.
■ *Earmarking:* Dedicating tax revenues to services (a form of user charges) reduces budget flexibility and reallocation during times of changing priorities.
■ *Coordination:* Managing individual services with dedicated user charges inhibits the coordination of public services.

Revenue by Infrastructure and Utility Sector

The Built Environment

Most of the built environment is in private ownership. The portion in public ownership is usually financed from current revenues and taxes. Bonds may be used to finance construction, with payback from revenues over a stream of time. There is no reason that there cannot be a charge for the use of a public building to recover its full cost, but if it is leased to private sector firms, then it appears that government is competing with the private sector. If a public building is owned by one agency and used by another, charges can be levied just as if the second agency had leased from a private owner. This results in funds from one government stream (such as a tax) going to another government agency (which might be repaying bonds).

Transportation

Basic transportation finance was discussed in Chapter 5, which explained the Highway Trust Fund and sources of revenue for other modes. Many state governments are mainly financed from income taxes, and they might combine these and revenues from state gasoline taxes with federal revenues to build highways. Once the highways are built, they might also be maintained and renewed from the same sources. These sources include, of course, other user charges from truck taxes. Transportation fees and charges focus on tolls, which can be used to charge for the use of roadways, transit fees, and transportation utility charges.

Energy Production and Distribution

Among the utility categories, energy use is the most straightforward to measure and ration. Therefore, among all public services, energy is the closest to a true utility. User charges can be used for operating expenses or to retire debt.

Electricity can be sold on a retail or wholesale basis by the kilowatt-hour (kWh), which is a unit of energy, whereas a kilowatt (Kw) is a unit of power, or the rate at which energy is produced. In Fort Collins, for example, electric customers can choose among the "energy rate" and the "demand rate." With the demand rate, users pay a charge for the rate at which electricity is used, based on the measured highest average demand for any 15-minute period during the billing month. The energy rate is better for low-demand users (less than about 1,400 kWh per month). Users can also select wind energy for an additional $0.01 per kWh. See Table 12-4 for an example of the city's charges.

According to the Fort Collins utility, kW demand and kWh usage are related but do not correspond directly. The kW demand is the rate at which electricity is used and reflects the capacity the utility must have. Demand is like the speedometer on a car, and use is like the odometer. Demand is averaged over time periods of 15 or 60 minutes, depending on customer class, and the charge is based on your highest average. For billing purposes, demand in kW is reset to zero at the beginning of each billing cycle (personal

TABLE 12-4. Comparison of energy and demand rates for Fort Collins

Measure	Energy rate, $	Demand rate, $
Fixed charge	3.91	6.44
Use charge per kilowatt-hour	0.0636	0.03031
Demand charge per kilowatt	Not applicable	4.1764

Source: City of Fort Collins 2007.

communication with Sharon Held, senior key accounts representative, Fort Collins Utilities, March 2007).

My own residential bill is on the energy rate. So, for example, on one bill, my monthly use was 674 kWh at a rate per kwH of $0.0694, for a total bill of $46.78. The fixed charge is blended with the use charge to increase the use charge per kWh slightly.

Natural gas service in Fort Collins is from Xcel Energy, operating through its subsidiary, the Public Service Company of Colorado. Xcel is a private business, operating as a public company and listed on the New York Stock Exchange. It is regulated by the Colorado Public Service Commission for service in the state. Figure 12-3 shows an example of an Xcel gas bill from a winter month in my own home. In the figure, service and facility pay for infrastructure, which includes pipes, compressors, and storage facilities. The franchise fee is paid to the city for the use of its rights of way. The voluntary energy outreach fund pays for people who are unable to pay their bills. The therm usage this month, 310, was for an average daily temperature of 22 °F. The bill also shows last year's same-month usage of 251 therms for an average daily temperature of 34 °F. (A therm is 100,000 British thermal units, and the multiplier adjusts it for altitude, temperature, and energy content.)

Rate class	RG-T Residential
Meter readings and usage, 27 days	561 − 245 = 316
Therms used w/ multiplier = 0.9796	310
Therms used same month last year	251
Usage charge (310 * .0802)	24.96
Interstate pipeline (310 * .06080)	18.85
Natural gas − Feb (126.15 * .6550)	82.63
Natural gas − Jan (183.85 * .5949)	109.37
Service and facility, franchise fee, sales tax	20.61
Energy outreach fund (voluntary)	5.00
Total bill ($9.49/day)	$256.32

FIGURE 12-3. Sample monthly bill for natural gas service in Fort Collins

Water Supply, Wastewater, and Stormwater

Water Supply Rates

Water rates illustrate the general principles behind rate setting, as explained in the American Water Works Association's (2000) basic manual on rate setting, *Principles of Water Rates, Fees, and Charges*. The rate-setting process for water supply begins with the determination of revenue requirements, the determination of the cost of service by customer classes, and the design of the rate structure. The association specifies two approaches to determine cost of service by classifying the costs differently, according to the commodity-demand method and the base–extra capacity method.

"Cost of service" is the traditional way to set rates at the marginal cost of providing service. The theory is that efficiency in the allocation of resources occurs when rates are set at the marginal cost of providing the service. Marginal cost means the cost to add a new unit of capacity. For example, if you are providing 10 million gallons per day of treated water and need to add 2 million gallons per day of capacity at a cost of $3 million per year, the marginal cost is $4.11 per thousand gallons delivered. The investment to achieve the current capacity of 10 million gallons per day might be less per unit of capacity, but marginal-cost pricing has you charge out the new capacity at its full cost. How to do this on an equitable basis is tricky, because you must distinguish between old customers who qualify for the old rate and new customers who pay the higher rate.

In reality, the cost to develop a supply may not include all the externalities, such as social equity and environmental quality, and a rate-setting body, such as a city government, may decide to set rates differently to recognize those. This is the basis, for example, for the "conservation rate" used by some water utilities. Figure 12-4 shows a sample of a local bill showing consolidated water, wastewater, and stormwater charges for Fort Collins.

The water rate in Fort Collins is tiered, meaning that the base charge is in place regardless of use. Then, for example, the first block of 7,000 gallons

Service	Usage	Unit charge	Total
Water	18,400 Gallons	2.83	52.13
Wastewater	5,452 WQA		19.23
Stormwater			16.21
Electric energy	674 KWh	.0694	46.78
City sales tax			1.32
Total			$135.67

FIGURE 12-4. Sample of a consolidated bill for water, wastewater, and stormwater charges in Fort Collins

goes at rate of $1.87 per thousand gallons (TG). The next block of 6,000 gallons goes at $2.15/TG, and the rest at $2.48/TG. The idea is that the "increasing block rate" will encourage people to conserve. Here is a breakdown of an example monthly bill:

Base charge	12.72
Block 1, 7,000 gallons, at $1.87/TG	13.11
Block 2, 6,000 gallons, at $2.15/TG	12.92
Block 3, 5,400 gallons, at $2.48/TG	13.38
Total of 18,400 gallons	52.13
Average $/1,000 gallons	2.83

For this two-person household during a 30-day month, the water usage was 307 gallons per capita per day, which seems quite large—except that it was during a summer month with lawn irrigation. If the in-house use was 100 gallons per capita per day, and if the lawn area was 5,000 square feet, the applied water use during the month was about 2 inches.

Wastewater Rates

Procedures for setting wastewater rates are not as well established as those for water supply. In the past, wastewater rates were set by splitting the cost among property taxes and user fees. Then, after the Construction Grants Program was initiated, the U.S. Environmental Protection Agency required user charge systems to be in place before awarding a grant. Now, wastewater has moved closer to an actual utility basis, whereby rates can be set according to discharge, that is, a more complete approach to the principle that the user pays.

In Fort Collins, wastewater charges are based on winter quarter average water usage (January–March). The numbers in the figures above do not tally because they come from different years. But to compute the charge, simply average water use rates for the winter quarter and apply the wastewater charge per thousand gallons.

Stormwater Rates

Stormwater is the newest water service with user charges using the utility concept to enable a city to collect operating and capital fees. This concept of charging for stormwater service as a utility rather than just from general taxes goes back to the 1980s. Cyre (1982) was one person who introduced the concept, and a number of cities have now tried it.

In Fort Collins, the city developed a stormwater utility in the 1980s. Charges are divided into development charges and monthly fees. The plant investment fee, which was initially the "basin fee," covers the infrastructure cost. It is applied to any new impervious surface of more that 350 square feet. As on 2007, the current charge was $3,070 per gross acre for all areas of the city, and this formula was used to calculate the PIF:

$$PIF = (\text{gross acres of development}) \times (\$3,070/\text{base rate per acre}) \times (\text{runoff coefficient})$$

Where:

■ Gross area in acres of each parcel of land includes open space and right of way.
■ The base rate is $3,070 per acre.
■ The runoff coefficient is determined by the percentage of impervious area in the development.

These runoff coefficients are used unless a development varies from the average: residential, 0.5, and commercial, 0.8.

The city also collects a development review charge. The stormwater base rate is $1,045 per net acre, computed by this formula:

$$\text{Development review charge} = (\text{net acres of development}) \times (\$1,045/\text{acre}) \times (\text{rate factor})$$

Where:

■ The net area in acres of each parcel of land includes open space (exclusive of right of way).
■ The base rate is $1,045 per acre.
■ The rate factor is based on a percentage of the impervious area and varies from 0.25 to 0.95 for very light to very heavy development.

Monthly stormwater rates are levied on all developed properties to pay for the construction and maintenance of the stormwater system. Rates are based on lot size (lot area plus share of open space in a development); a base rate of $0.0041454 in the city; and a rate factor, which is based on the impervious area.

A calculation for a single-family property would be (City of Fort Collins 2007):

$$\text{Lot size} \times \$0.0041454 \times \text{rate factor} = \text{monthly rate}$$

Any square footage over 12,000 pays only 25% of the fee.

Solid Wastes
Solid waste collection services are readily privatized and can be financed entirely by user fees. Recycling plants and landfills can be privatized as well,

but private collection services are more common. Transfer stations, incinerators, and resource recovery plants are also used in solid waste management.

The City of Fort Collins mandates charges by private collectors and operates the landfill and recycling center. Though the city does not operate a solid waste system, it works to promote recycling. In 2006, it commissioned a study on how to encourage recycling (Skumatz Economic Research Inc. 2006). Many other cities continue to operate solid waste management systems.

Paying for System Expansion and Renewal

Infrastructure is required for new or expanded facilities or to renew old ones. Either way, financing decisions are about why or whether to spend, how much, who pays, and how to structure the financing. These decisions involve different financing studies, such as:

Why or whether to spend: justification/planning
How much to spend: optimization/budgeting
Who pays for infrastructure: cost allocation
How to structure the financing: financing mechanisms

Justification and planning studies require financial or economic analysis using benefit-cost or rate-of-return techniques. At the federal level, studies to justify public investments usually analyze whether the project's benefits exceed its costs (see Chapter 6 for an explanation of benefit-cost analysis). The idea is that an investment is justified regardless of who gets the benefits. At the state and local levels, this concept is not used as much because the analysis usually focuses on filling a need and determining how to pay for it. If private funds are to be invested, the analysis will study whether the return to investors exceeds a minimum attractive rate of return.

The amount of capital to invest is limited, so the investor must allocate available capital over the possible investments to maximize the rate of return while managing risk. So, for both private and government investors, the questions are about investment strategy.

On the private side, optimization is a balance between return and risk. Investing in one venture may have the potential for the greatest return, but the investor may want to diversify to manage risk and balance a portfolio through asset allocation. If the investor is a unit of government, it is impossible to determine an optimum social rate of return for a portfolio of investments, so capital funds are allocated according to political objectives as well by using social and economic analysis. As an example, to get a spending bill through Congress, funding has to be spread around to get enough votes to pass. In a city, it will be necessary to have capital projects in different parts of the town and to benefit different users or the program will not be approved. Economists refer to maximizing return as the size of the pie and to the equity part as the size of the pieces.

The capital financing of infrastructure requires a basic decision about "who pays?" Either the people who receive the benefit from the infrastructure pay or everyone pays into a common pool. A principle of utility financing is that those who benefit should pay. To implement this principle requires identifying classes of people who should pay, including not only where they live and what they do but also when they use the benefits of the service. Classes of people to pay for services include, for example, residential customers and commercial customers. The classes of payers could involve people driving personal vehicles versus truckers. When it is not possible to identify the beneficiaries, the services are pooled and paid from general revenues.

The "when" question raises the issue of scheduling the financing package. If a town has its infrastructure paid for and no expansion is required, the answer is easy: It pays for renewal and modernization. If a town is built from scratch, the answer is also easy: The new people pay to build the infrastructure. If a town has a combination of current and newly arriving residents, the answer is more difficult. This is the origin of the debate over how "growth pays its own way." The city's accounting system has to be good enough to compute accurately the costs required by renewal and expansion. The other issue looks at the time dimension. Do people already living in the town contribute to a fund to build in the future, or do you borrow and build it now, paying later from revenues contributed mostly by new residents? This is the difference between "pay as you go" (current revenues) and "pay as you use" (debt) financing.

Figuring out how much each class of customer pays is a "cost allocation" exercise and the central task of a rate study for a utility. Figuring out which customers pay in terms of when they use the service is the essence of the debate about debt versus current revenue. That is, if current revenue is used, today's customers pay. If debt is used, tomorrow's customers pay. In the federal budget, this latter issue is sometimes called "intergenerational equity." Must tomorrow's workers pay for Social Security for today's workers when they retire?

Capital Strategies

In planning a capital strategy, the infrastructure manager will look at the mixture of payers and beneficiaries. It may never be possible to determine exactly how much to invest and who should pay for each detail, but in the real world a package is assembled and presented to the approval authority. The package may comprise debt, fees, new taxes, and other mechanisms. Projects can be mixed and matched to create a feasible package that has enough appeal to be approved.

Pay-as-you-go financing with current revenues is like buying a new car with cash. You have saved as you go along, and when it's time to buy, you pay cash. In infrastructure, when financing is from current revenues, the main sources are fees and taxes. In pay-as-you-go financing, funds go into a capital reserve account, which is earmarked for use when needed.

Current revenues are popular for infrastructure financing and are the easiest way to finance projects due to administrative ease and lack of carrying charges. They are also easily understood by the public and politically acceptable. However, current revenues are easy to divert from capital spending when other priorities hit or when a crisis occurs, as they inevitably do. Also, the use of current revenues compels today's residents to pay for infrastructure to be used in the future. Going back to the car purchase example, it's like you've been paying into a new car fund, but someone else gets to drive the car when it's finally purchased.

To overcome problems with pay-as-you-go financing, debt (or pay-as-you-use) financing can be used. If the term of the debt repayment is the same as the life of the facility, then the amortization of principal and interest in regular payments means that the facility is paid for just as it needs replacing. This involves two "lives" of facilities; one is the service life (how long it actually lasts), and the other involves the time to pay for it.

Of course, the service life is uncertain because we do not know, in general, the lifetimes of facilities. However, we can compute life cycle costs with some accuracy. This is because as service lives get longer—as, for example, a 100-year-old cast iron pipe, the capital cost diminishes in comparison with the operation and maintenance costs.

In spite of administrative expenses, debt financing can be cost-effective due to inflation, uncertainty, and the opportunity to invest revenues elsewhere (known as "arbitrage," or borrowing money in one place and then investing it somewhere else).

If interest rates on debt are low, then it pays to borrow and build now because it might cost more in the future. If interest rates are high, current revenues might be a better choice now, with the option of borrowing at more favorable terms later.

Debt financing works well with enterprise management, especially when revenue bonds are used. By the 1980s, revenue debt was nearly three times general obligation debt, but general obligation bonds remain important to infrastructure (Vaughn 1983; Valente 1986). Just as there are limits to personal debt, there are limits to bond financing, usually based on a ratio of bond indebtedness to the assessed valuation or revenues of an enterprise.

System Development Charges

System development charges are also used to finance infrastructure. Other names for these charges are impact fees, development fees, plant investment fees, and tap fees. In a system development charge, you compute the capitalized cost of an element of infrastructure required to provide a public service. This could be the generation capacity for electric power, the cost of school buildings, the cost of roads, or a number of other similar measures. The idea is that people who connect to the system as users pay a share of this cost. It's the same principle in many ways as joining a club where your initiation fee buys a share of the enterprise.

System development charges became popular during the 1970s as a way to have "growth pay its own way." They were used more where there was much growth, as in the West and in Florida. A survey taken in 2002 showed that they had not spread as much as might be anticipated. Use declined from 36% to 25% of municipalities surveyed. In the Pacific Coast region, 62% used them. In the Mountain West, 55% used them. In the East, mainly unincorporated areas in Florida used them (Lawton 2003).

System development charges allow new users to "buy into" an existing system by paying their fair share of it. Consider a community with a water supply system with capacity for new developments. A new development is charged a system development fee to pay for its share of the system. This fee would be passed on to the purchaser of the developed property in the form of higher costs for their land or as fee itself. Ultimately, it passes the cost of the infrastructure on to the property owner.

Fort Collins uses the principle of "growth paying its own way" to justify system development charges for new developments. The ones currently in use are water plant investment fees, water rights acquisition charges, sewer plant investment charges, storm drainage fees, street oversizing fees, off-site street improvements, electric offsite and onsite fees, and parkland fees. As you can imagine, these fees add substantially to the cost of a home.

Grants

Grants, or "intergovernmental revenue," are an important part of the financing of local infrastructure. They transfer capital from one group to another. The largest share of grants is from the federal government, which gets its revenue from various tax streams, including individual income taxes. Thus a grant from the federal government to a city means that people all over the United States are paying for something in that city.

The people in the city also pay taxes, so another way to look at it is that they are getting their tax money back in the form of a grant. The problem is that the federal government keeps part of their tax money and doesn't send all of it back. Here you have a reason people don't like taxes: They don't get to decide about how their money is spent, because the decisions are being made by government officials.

The U.S. budget contains funds for intergovernmental grants. Chapter 11 explains the budget in detail. State governments also provide grants to local communities.

Tax Increment Financing

A variation of debt financing that also involves public private cooperation is "tax increment financing," defined as an approach that uses the increase in taxes that occurs after a development is finished to repay debt. An example of this occurred in Fort Collins when voters approved a proposal to build downtown redevelopment facilities using funds raised by the Downtown Development Authority, which was to sell tax exempt bonds to reimburse developers for some of their expense for a senior citizen's housing facility to be built in conjunction with office and retail space. This program has been successful, and a state bill to extend its authorization was passed in 2008.

Revolving Funds

Revolving funds are a form of development bank. Examples are a clean water revolving fund and a drinking water revolving fund. Such a fund is established through a capital infusion, for example, from the U.S. Environmental Protection Agency's Clean Water State Revolving Fund account established to finance wastewater treatment. Congress appropriates funds for the program, and these become the capital to establish the fund. States operate the fund, and local governments or wastewater utilities borrow from it, repaying with interest.

Challenges in Developing Countries

Setting user fees may be difficult in developing countries where incomes are not high enough to support the infrastructure. People with subsistence-level incomes simply cannot afford expensive infrastructure systems, and if these systems are necessary for social and economic development, some interim means to finance them is necessary so people can get on their feet and develop the capacity to pay. This is a chicken-and-egg problem in the sense that they need the infrastructure to prosper but cannot pay for it until they prosper.

Applications to Infrastructure and the Environment

Managers agree that coming up with sufficient funds for their operating costs and investment needs is a paramount problem for infrastructure and environmental systems. A general revenue model shows these managers how the operating and capital costs of capital facilities and public utility services are funded. Revenues include taxes, rates, charges and fees, and plant investment charges.

A capital strategy shows infrastructure and environmental managers how to use combinations of pay-as-you-go and pay-as-you-use financing. This leads to a knowledge of debt financing and the bond markets, as well as the use of capital windows such as revolving funds. Knowing about revenue models also alerts the manager to the possibility that funds could be lost through mechanisms such as administrative fees or requirements to make a payment in lieu of taxes.

References

American Water Works Association. (2000). *Principles of water rates, fees, and charges, manual M1*. 5th ed. American Water Works Association, Denver, CO.

City of Fort Collins. (2007). Stormwater rates. http://www.ci.fort-collins.co.us/utilities/rates-stormwater.php. Accessed July 21, 2007.

Colorado Division of Property Taxation. (2004). *Understanding property taxes in Colorado*. Colorado Division of Property Taxation, Denver.

Cyre, H. J. (1982). Stormwater management financing. Paper presented at American Public Works Association Congress, September, Houston.

Lang, Jennifer. (2007). "New urbanism's flip side." *Rocky Mountain News*, February 24.

Lawton, L. (2003). *Development impact fee use by local governments, municipal yearbook*. International City/County Management Association, Washington, DC.

Skumatz Economic Research Inc. (2006). *Fort Collins solid waste 5-year strategic plan: Strategies to reach 50% diversion from landfill disposal*. City of Fort Collins, CO.

U.S. Bureau of the Census. (2007). State government tax collections, 2005. http://www.census.gov/govs/statetax/0534ncstax.html. Accessed February 18, 2007.

U.S. Office of Management and Budget. (2007). Federal receipts and collections: 2008 budget request. http://www.whitehouse.gov/omb/budget/fy2008/pdf/apers/receipts.pdf. Accessed February 22, 2007.

Valente, M. G. (1986). *Local government capital spending: Options and decisions, municipal yearbook.* International City/County Management Association, Washington, DC.

Vaughan, R. J. (1983). *Rebuilding America, volume 2: Financing public works in the 1980s.* Council of State Planning Agencies, Washington, DC.

13

Capital Financing and Markets

Capital Financing for Infrastructure Systems

Some infrastructure systems, such as roads and water distribution, have large percentages of fixed assets and are capital intensive. Others, such as a bus transit agency, require higher ratios of operating funds. Regardless of these ratios or the funding sources, infrastructure systems require large amounts of capital for construction and renewal, as well as to acquire equipment and other fixed assets.

As we saw in Chapter 2, all fixed assets in the United States are valued at about $40 trillion. Nearly four-fifths of the value is in private residential and nonresidential structures and equipment, and with the downturn in housing prices, this part will decline in value. The other fifth is in public infrastructure and equipment, which constitute the capital base for public infrastructure systems. Of this, most comprises structures and government capital stock, such as roads and bridges.

Whether on the private or public side, funds to finance infrastructure systems are usually obtained through debt financing in the capital markets. The capital markets are sources of financing from stock or bond investors, banks, loan funds, and any other source of credit that enables organizations and individuals to build, expand, and renew their physical assets.

Capital for construction can be provided from current revenues, debt financing, and/or intergovernmental subsidies. Current revenues from service or impact fees or from taxes for capital construction are called "pay-as-you-go" financing. Debt financing from loans or bonds is called "pay-as-you-use" financing. Intergovernmental subsidies are mainly through grants from another level of government, such as federal grants to cities, which, as

Chapter 11 explained, have been declining. Chapter 7 explained how current revenues can be used for pay-as-you-go financing or to service debt for capital financing. This chapter focuses on debt financing and explains how the capital markets work as sources for investment funds for infrastructure.

The chapter explains sources of capital financing, including the bond markets and other sources of capital. It does not claim to present a complete explanation of these gigantic and complex markets. Rather, the scope of the chapter is limited to explanations of how the capital markets work and how they are used to finance infrastructure in both the public and private arenas. Environmental finance is part of infrastructure finance, except for the parts of it devoted to funding regulatory programs and preservation through land acquisition.

How Credit and Debt Financing Evolved

In today's capital markets, infrastructure agencies can raise the funds they need if they have an adequate revenue base to repay their debts. Providing credit to these agencies requires decisions by those who have capital to lend it in exchange for promises to repay so they can earn fair returns on their capital.

People have needed credit from early recorded history. The Old Testament warns against "usury," which is understood to mean exorbitant rates of interest. However, few people today deny that a lender is entitled to a fair return on its capital. Before the days of modern corporations, farmers and small businesses needed loans to buy seeds and equipment and launch enterprises. Today, giant enterprises need access to borrowed capital to build their facilities and to finance operations.[1] Even the U.S. government borrows capital for various purposes, such as to fight wars.

Although credit is available on a widespread basis in today's global capital markets and international finance system, it has not always been this way. Today's capital markets emerged mostly in the twentieth century, and the most rapid changes have been in the last two decades.

In capital markets, access to credit requires organized lending systems. The most basic form of these is through banks, and from its founding, the United States began to develop its banking system. Chapter 2 briefly recounted this history. Early on, the money and banking system was not as sophisticated as it is today, and the only way you could get a loan was from an individual or a small, independent bank. As the United States' fledgling banking system developed, the money supply was limited by the supply of

---

[1]Infrastructure enterprises normally do not borrow capital to finance operations, which should be funded from current revenues. See Chapter 11 about the rules for government finance.

gold because each dollar had to be backed by gold. This meant that there was only so much money to lend. As a result, small farmers and businesses had a hard time accessing capital to expand and operate. By the end of the nineteenth century, this was a major social problem because farmers, small businessmen, and craftsmen did not want Wall Street bankers and other capitalists to profit from tight money policies based on the gold standard that blocked their access to credit. Advocacy of "bimetallism" meant to use silver as well as gold as a basis of money and therefore to ease the supply of credit. President William McKinley and later presidents looked for ways to increase the supply of credit by easing the money supply.

Today, credit is available through many channels. In addition to a more robust supply of credit for everyone, the field of public finance developed during the last century and provided mechanisms for the financing of public infrastructure, including tax-exempt bonds. Given the problems in the credit markets with the giant financial crisis, only time will tell how the nation resolves future credit issues.

The Infrastructure Sector's Financing Needs

Financing for infrastructure must compete with other calls on investment capital. As Chapter 2 explained, this investment capital is needed to finance both private and public infrastructure, businesses, and consumer goods.

In Chapter 2, we showed how the capital stock of the United States was about $30 trillion in 2000 and had grown to about $36 trillion by 2004. Capital stock is a measure of the total value of public and private infrastructure. The distribution of the stock by categories in 2000 figures is shown in Table 13-1, which illustrates the relative percentages. The numbers were relatively stable until the large decline in home values beginning in 2007 or so. Therefore, when new figures are released, they will show shifts in the percentages.

We can get a perspective on the $5.7 trillion government stock by looking at some approximate estimates. The United States has some 4 million miles of road, and if they were valued at $1 million per mile, this would reach $4 trillion. It has about 1 million miles each of water and sewer pipe, and if the value per mile was $0.5 million, this would reach $2 trillion. Adding these two reaches $6 trillion. Add in all the buildings, equipment, energy facilities, dams, and related public infrastructure, and the figure goes higher. It is probably not reasonable to value each mile of road at $1 million or each mile of pipe at $0.5 million, and this is only an order-of-magnitude estimate.

To see the big picture of national assets and the need to finance them, notice how Table 13-1 shows that private stock dwarfs government stock.

TABLE 13-1. Public and private capital stock in the United States, 2000

Capital stock	Share of total (percent)	Value (trillions of $)
Government		5.7
Federal	25	
State and local	75	
Structures	88	
Equipment	12	
Private		21.1
Private nonresidential equipment		4.2
Private nonresidential structures		6.4
Private residential housing		10.4
Net government and private		29.6

Source: U.S. Bureau of the Census 2007.

Chapter 2 also showed that financial assets are much greater than nonfinancial assets. Thus, the infrastructure to be managed makes up a relative small share of total assets but still totals more than $5 trillion nationally, and when private infrastructure is added, the figure is considerably higher. In a later section on the bond market, we compare the total value of public capital stock to outstanding bond indebtedness, which are important indicators of how much was borrowed to finance that stock.

Capital Markets

Roles of Capital Markets

Capital markets are places where capital is exchanged from those that have it to those that need it. The main sources of capital are banks, the stock market, the bond market, and private equity sources. Banks come in many forms and operate according to the money and banking system and its regulatory controls. In the stock market, businesses raise capital in exchange for ownership shares, whereas in the bond market, businesses and government raise capital in exchange for their promises to repay. Private equity, which is becoming a significant source of capital for revenue-producing infrastructure, matches private lenders and investors with opportunities in the infrastructure sector. For public sector debt, the main source of capital is the bond market.

Capital is wealth that has been stockpiled and is available for lending or use. Wealth is derived from savings or capital formation in the form of property, money in storage, or business equity. The money in storage can be used to finance the labor, energy, and materials that are required to construct infrastructure (as well as to invest in businesses and other productive enterprises).

As you can see in Fig. 13-1, capital funding in the form of loans or bonds is transferred to enterprises that build facilities and deliver services to users. The users pay fees or charges to the enterprises, which in turn pay back the loans or bonds with interest. If a default occurs, the capital markets suffer a risk loss, which is part of the cost of doing business.

Capital Institutions

Capital market institutions that serve the infrastructure and environmental sectors include banks, stock and bond markets, and other investment and credit sources. Commercial banks are ready to lend money to businesses and nonprofit enterprises, but they lend money less often to government agencies. This is a reason that development banks and authorities have been created in states and countries where infrastructure investment is required. See Chapter 2 for a discussion of types of banks. Lending money is not limited to banks, as it was in the past. Other investors, such as insurance companies and pension funds, will invest in solid enterprises that offer them attractive returns.

The stock market is a place where capital can be raised by private companies, many of which participate in infrastructure enterprises or related businesses. How the stock market operates is explained later in the chapter.

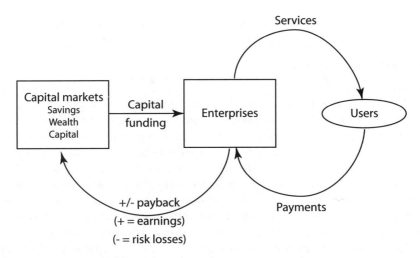

FIGURE 13-1. Capital markets for infrastructure projects

It focuses on investment capital for private businesses, not government agencies.

The bond market is also explained later in the chapter. An example of its use to raise capital might be a toll road authority that required construction funding from a bond issue to add lanes and renew its road system. After the necessary engineering studies, it would normally utilize an investment bank to serve as underwriter, to issue the bonds, and to serve as the trustee.

As investment vehicles, stocks and bonds are fundamentally different. Stock represents ownership in companies, and stocks are therefore called "equities." Bonds are promises to pay interest and return the principal to the bond holder at a later date. To an investor, stocks and bonds represent ways to earn interest and/or capital gains on money. To those needing capital, bonds are a way to borrow money and stocks are a way to sell ownership in an enterprise in exchange for capital.

Agencies with authority to lend money are operating like development banks. Normally, they will have access to an initial source of capital, and they will replenish their capital through loan repayments.

Private equity investors and venture capitalists represent privately owned capital sources that will take an equity stake in an enterprise in exchange for capital. Whereas we associate these more with startups and buyouts than with capitalizing infrastructure ventures, utilities or other revenue-producing enterprises may be attractive targets for private equity takeovers.

Stock Markets

The stock market includes infrastructure businesses and companies with large business activities in the environmental and natural resources fields.

Capital for Business—Emergence of the Stock Market

Stocks and the stock market fill the basic need of providing ownership in enterprises in exchange for invested capital. Today's system of public ownership of stock in companies traces its roots back centuries, when money was needed to finance ventures of different kinds. Belgium had the world's first stock exchange in 1531, and stock shares to finance the East India Company were traded in 1602 in Amsterdam. London had an informal stock exchange by the 1700s, and Wall Street became the center for trading in the United States in 1792. This led to the establishment of the New York Stock Exchange in 1863 and to its later competitors, such as the American Stock Exchange. By 1900, the stocks of a number of large companies were being traded, including the then-hot railroad stocks. The biggest stock in 1900 was

U.S. Steel, and other companies that emerged early included names that are still familiar, such as AT&T, Westinghouse, Kodak, Procter & Gamble, Pillsbury, Sears, Kellogg, and Nabisco (Stocks Investing 2006).

The history of the stock markets includes the achievements of famous investors such as today's Warren Buffet. His mentor was Benjamin Graham, known as the father of security analysis. Graham's classic book *Security Analysis* explains the popular notion of "value investing," or investing in companies that have significant underlying or basic value. Graham's (1996) story explains how he got his start on Wall Street, at a time when brokerage firms discriminated against Jews and Graham had a hard time finding employment. He overcame that stigma and, after graduating from New York's Columbia University, he taught there from 1928 to 1957 while working in the securities business.

About Stocks

Stock represents equity capital supplied by the owners of an organization. Issuance of stock is generally governed by state laws for corporations and regulated by state and federal law. For example, if you incorporate a consulting engineering company in your state, you will file incorporation papers with your secretary of state or the equivalent officer.

The main two classes of ownership are common stock and preferred stock. Common stock represents ownership in a corporation and carries the right to vote on the membership of the board of directors, which oversees the firm on behalf of the stockholders. Preferred stock does not represent ownership and does not usually carry voting rights, but it enjoys preference in the granting of dividends from net earnings. Regulated utilities often issue preferred stock, which provides a source of equity capital without going to the debt markets (Melicher and Norton 2005).

Stocks are generally valued according to the financial performance of the company. A publicly traded company paying a steady dividend of, say, $5 per share per year might be worth on the order of $100, if the fair return on equity capital was judged to be 5% and there was little growth potential in the company. If the company was earning that same $5 and paying all of it out as dividends, with no reinvestment or growth, then the earnings per share would be $5 and the price/earnings (P/E) ratio would be 20.

If, conversely, the same company was earning $5 per share but paying only $1 as dividend and investing the rest for growth, the market might bid the price of the stock higher on its growth potential. The P/E ratio might be 40, and the stock could be valued at $200 per share.

Stock trading is done to maximize the potential for income and capital gains. For example, if you buy a share of stock in XYZ Corporation today at $50 and sell it in a year at $60, you have earned a capital gain of

$10, or 20% in a year. You might also receive a dividend distribution of, say, $5, thus bringing your gain in a year to $15, for a rate of return of 30%, before taxes.

The Use of Stock in Privately Held Corporations

Sometimes small businesses, such as consulting engineering or construction companies, issue corporate stock as a way to share ownership among key employees. They may not raise capital directly from the stock, but they may be raising capital indirectly because those key employees may give loyalty and extra effort in exchange for the stock. In the same way, stock can be used to reward employees, such as through stock options or distributions to key employees.

Today's Stock Markets

Stock markets are places to trade publicly owned shares. These can be formal places, as in the case of stock exchanges, or less formal, as in the case of over-the-counter markets or privately brokered stock transactions. A number of construction industry companies, utilities, infrastructure industry suppliers, and natural resources companies are listed on stock exchanges.

In the United States, the principal stock exchanges are the New York Stock Exchange (NYSE) and NASDAQ, which is the National Association of Securities Dealers Automated Quotations system. The American Stock Exchange has merged with NASDAQ.

The NYSE traces its history back to 1792, when stockbrokers on Wall Street organized themselves. The name NYSE dates to 1863. The current building at 18 Broad Street was finished in 1903 and was listed as a National Historic Landmark in 1978. It is the largest stock exchange in the world by dollar volume, and it had a market capitalization of around $25 trillion in 2006. NASDAQ lists more stocks, however (Kansas 2005).

Stock trading and the brokerage profession have changed dramatically during the last few decades. The rise of the mutual fund industry and its variants and online trading have changed the past arrangements whereby you might order transactions only through a broker, who had access to a seat on the stock exchange. With Internet trading, a global stock market is rapidly becoming a reality. For example, the NYSE and Euronext have merged to form a transatlantic stock and derivatives exchange, known as NYSE Euronext (Lucchetti et al. 2006). Other stock market mergers are in the works.

The decline in the stock markets during the period 2007–9 was the most dramatic fall in decades, and only time will tell how long it takes to recover.

Given this occurrence, the nation will also see an institutional shakeout in brokerage firms and others involved in the capital markets.

Stock Market Indices

Given their large financial stakes, stock markets operate with a vast amount of rapidly changing information. The most important information is the price of a stock at a point in time, which some people believe incorporates all the available information and is the best indicator of the stock's value. The prices of stocks within sectors are tracked through a number of stock market indices, such as the Dow Jones Industrial Average, sometimes called "the Dow." This long-standing index is basically the adjusted value of a sum of 30 major stocks. The Dow purports to represent the market on a broad basis, but its inclusion of major stocks only gives it a bias toward "large cap" stocks, or those with large market capitalization.

Other indices capture different aspects of the market. For example, the Standard & Poor's 500 index is a broader market index. The NASDAQ composite measures high-technology stocks and captures that stock exchange's stocks and the Russell 2000, which lists some 2,000 small cap stocks. There are, of course, many other stock indices as well.

Indices of stock prices within sectors of the economy are also recorded, such as the Dow Jones Transportation Index and Utility Index. The utility index for the most part consists of electric power utilities, and the transportation index measures airlines, railways, and shipping companies.

A mutual fund aggregates funds from many investors and creates a portfolio of investments, usually targeted toward a fixed range of investment objectives. For example, a fund could be a "value" fund that looks for lower-cost stocks that might have been overlooked. Or it could be a "large cap" fund that mainly invests in large companies.

Commodities and Futures Markets

Another venue for investments is the set of commodity and futures markets, such as the Chicago Mercantile Exchange. In commodity markets, raw or primary products are bought and sold through contracts. The products can include physical products such as food, metals, or electricity. Agricultural products such as grain and livestock have been traded for decades, and new commodities are added from time to time. Commodity and futures contracts are based on agreements to buy now and pay and deliver later. Today, what were known as forward contracts have become known as futures contracts. Hedging is made possible through futures contracts and can insure against poor harvests by purchasing futures contracts in the same commodity that a farmer grows.

Many commodity markets started in the late twentieth century. Oil trading is now of great interest. Environmental capital has also entered the world of trading. Emissions and weather trading are other examples of "negative commodities." Hedging can be used to avoid the consequences of damage from natural disasters through weather "derivatives," which deal with issues such as the impact of drought or frost on crops.

Debt Financing Through the Bond Market

The Evolution of Bond Markets

Bond markets are nearly as old as stock markets. State loan stocks were used as far back as the fifteenth century (Brown 2006). All types of bonds involve borrowing with a promise to repay with interest. War bonds were used to finance World War II, and today the U.S. national debt is guaranteed by government Treasury bonds and bills. Today, the bond market is a major source of capital financing for many purposes. In 2006, the total bond indebtedness of the United States was almost $26 trillion, or about $87,000 per capita, with this breakdown (Bond Market Association 2006):

■ mortgage related (agencies such as Ginnie Mae, Fannie Mae, and others): $6,095 billion,
■ corporate debt: $5,095 billion,
■ Treasury backed (interest-bearing marketable public debt): $4,322 billion,
■ money markets: $3,497 billion,
■ federal agency securities: $2,641 billion,
■ municipal bonds: $2,256 billion, and
■ asset backed (backed by home equity loans, credit card receivables, auto loans, etc.): $1,966 billion.

As you can see, the greatest debt is mortgage related, followed by corporate debt, then Treasury debt, then money market, and federal agency securities. Municipal bonds, with which we are most concerned here, are still $2.3 trillion.

Tax-Exempt Municipal Bonds

Municipal bonds are tax exempt, representing a subsidy that has been around since an 1895 Supreme Court case (codified in the Revenue Act of 1913) (U.S. Internal Revenue Service 2006). The tax-exempt feature of municipal bonds represents a federal subsidy for infrastructure construction because the government does not derive any tax revenue from interest received by investors. Conversely, the bonds carry a lower interest rate because they are tax exempt. This lower interest rate wipes out some of the incentive to invest in municipal bonds, but the reliable nature of most infrastructure organizations adds to their attractiveness.

Tax-exempt bonds may be issued as governmental or nongovernmental (private activity) bonds. Governmental bonds are issued by state and local governments, for such projects as highways, government office buildings, and water and sewer facilities. Private activity bonds may be taxable. Examples include activities sanctioned by a government but carried out by a private entity, such as an industrial development project.

The main bonds that finance infrastructure are general obligation (GO) or revenue bonds, which are backed by the full faith and credit of the organization issuing the debt. They may be paid off from a source of revenue, but the guarantee is with the taxing power of the entity. An organization must have taxing power to issue GO bonds. GO bonds are logical for projects with community-wide benefits, such as municipal buildings, public schools, streets and bridges, and economic development programs.

Revenue bonds are repaid from the revenues of an enterprise such as a toll road or water supply system and are not guaranteed by a tax base. They can be issued by more entities than GO bonds, and they are viewed as riskier and thus have higher interest rates. Logical services for revenue bond financing are those that have user fees, such as water and wastewater, electric power, solid waste, parking garages, and airports.

Industrial development bonds are a type of private activity bond. For instance, after the terrorits attacks of September 11, 2001, Congress authorized the Liberty Bonds Program to provide tax-exempt redevelopment funds for Lower Manhattan's commercial market. The program, which was unusual in that it authorized bonds for development outside a blighted area, was authorized in 2002 and was scheduled to expire in 2006, when the $8 billion in bond funds was gone (Pristin 2006).

The Bond Issuance Process

When an infrastructure entity needs to raise funds from bonds, it determines how much money is needed and when, and then it turns to a bond house for advice on the best deal. The process of issuing bonds is shown in Fig. 13-2. The bonds are sold to finance a project that provides infrastructure for a utility or an authority that provides services such as water supplies or a toll road. The bonds are offered for sale by the authority, which works through an underwriter and trustee to sell them to the bondholders and manage the bond issuance process. Revenues flow back from the users to the issuer and eventually to the bondholders.

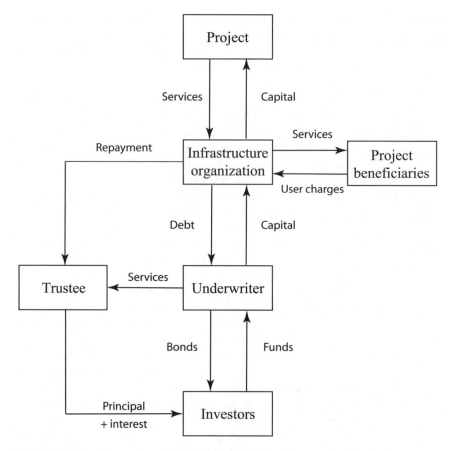

FIGURE 13-2.   Flow of activities in the bond issuance process

Bond issues are generally handled by underwriters or investment banks. You will find underwriters marketing their services at conventions of groups such as the Government Finance Officers Association, the National Association of State Treasurers, the Airport Operators' Council, the International City/County Management Association, and the International Bridge, Tunnel, and Turnpike Association, as well as other infrastructure professional associations. Investment banks handle more services than just bond issues. Public bonds might be one of the branches of their operations, which are explained later in the chapter.

Sometimes, an authority will choose to refund its bonds to use the proceeds to pay off old bonds to gain more favorable terms. Therefore, an infrastructure authority might be continually active in the bond market as it tries to minimize the cost of its debt issues.

Bond Risk

Bonds are rated according to risk, and infrastructure authorities seek the best ratings because if ratings are low, interest charges are higher to compensate investors for their risk. If interest charges are higher, then the infrastructure authorities must pass the charges on to customers in the form of higher rates or charges.

To assess risk, bonds are rated by rating agencies. The main three are Standard & Poor's Corporation, Moody's Investor's Service, and Fitch. A. M. Best and Dominion also rate investments, but they are smaller than the others.

People tend to think of utilities and infrastructure organizations as safe investments, but defaults can and do occur. In the 1970s, New York City had a financial crisis. It was financing ongoing operations with bonded indebtedness. This led to a reexamination of New York's financial policies and the creation of a state program called the Municipal Assistance Corporation. Also, in the 1990s, Orange County, California, faced a crisis brought on by a tax-limitation initiative in the state. It was also resolved but was another close call for municipal bonds (O'Higgins 2000).

A large bond default occurred in 1983 on $2.5 billion in bonds of the Washington Public Power Supply System, or "Whoops" bonds. These bonds had been used to finance nuclear power (Leigland 1986). And in 2006, the Eurotunnel authority threatened to go bankrupt unless it restructured its debt (Sakoui 2006).

It is interesting to compare the total value of public capital stock with outstanding bond indebtedness, which should be the principal indicators of how much was borrowed to finance that capital stock. I showed above that municipal bonds are at about $2.3 trillion, whereas the value of public capital stock, as a component of U.S. wealth, was about $5.7 trillion (in year 2000). Of this, some 75% is state and local capital stock, whose financing is often through bonds. The federal government does not issue municipal bonds because it can finance itself through Treasury bonds. So, in round numbers, state and local public capital stock is about $4.3 trillion and outstanding municipal bonds are about $2.3 trillion. This means that to pay for their existing infrastructure, the governments of the United States are about 53% in debt, and they have a lot of deferred maintenance and renewal as well.

If Interest Rates Go Up, the Prices of Bonds Go Down

Bonds are bought and sold and thus are not always held to maturity. An interesting feature of bonds as an investment is that if the interest rates in the economy go up, the prices of bonds tend to come down. This occurs because when bonds are sold initially, they have a fixed interest rate. Thus, if you pay the par value and keep the bond to maturity, you will receive the designated rate of interest throughout the life of the bond. However, if the original interest rate reflects the cost of money at the time the bond is issued, and if the prevailing interest rate then goes down, it makes the bond more valuable because it earns more interest. Going the other way, if the interest rate goes up, your bond is now worth less because it earns less than the prevailing rate of interest.

Investment Banking

Often, the firms that offer municipal financing services also engage in other forms of investment banking. The activities of investment banks include the issuance of securities (stocks and bonds), helping investors buy securities, managing financial assets, trading securities, and providing financial advice. Historically, well-known firms included Merrill Lynch, Salomon, Smith Barney, Morgan Stanley, Dean Witter, and Goldman Sachs. Some of these have changed dramatically during the financial crisis. For example, Merrill Lynch was acquired by the Bank of America in a merger forced by the federal government.

Some investment banks specialize and might be called boutiques that are oriented toward an activity like bond trading. As an example of the organization of an investment bank, Goldman Sachs is divided into investment banking, trading and principal investments, and asset management and securities services groups. Its Investment Banking Division handles mergers and acquisitions, divestitures, and the issuance of equity or debt capital. Its Trading and Principal Investments Group focuses on making money for the firm. And its Asset Management Group serves wealthy individuals by helping them manage their assets (Careers in Finance 2006; Goldman Sachs 2006).

Private Equity

Today, private equity sources have increased their influence in financing all sorts of capital expenditures as well as business enterprises. Private equity means any kind of ownership through securities not listed on a public exchange. Private equity firms make profits by buying a firm, issuing stock to the public, and selling or merging the company, or through some sort of recapitalization. Private equity funds aggregate contributions from smaller investors to create a capital pool that can be used for investment purposes, usually for institutional investors and wealthy individuals. In 2006 public pension funds, banks, and other financial institutions made 40% of all private equity commitments. A fund of funds is a private equity fund that invests in other private equity funds to lower risk exposure through diversification. These accounted for 14% of commitments to private equity funds in 2006. Venture capital is a type of private equity that focuses on investments in new and maturing companies. Private equity might get involved with infrastructure enterprises if it sees them as profitable as compared with other investments the fund might make (Maxwell 2007).

Capital Market Regulators

The stringent regulation of capital markets is necessary because of the possibility of fraud, given the large sums of money involved. The regulation of banks was discussed in Chapter 2. The stock markets and stock transfers of public companies are regulated by the Securities and Exchange Commission (SEC), whose mission is to "protect investors, maintain fair, orderly, and efficient markets, and facilitate capital formation." It regulates stock exchanges, brokers, advisors, and mutual funds. Much of its work comes through disclosure requirements. Public companies must disclose financial and other information to create transparent capital markets and to facilitate capital formation. Violations might include insider trading, accounting fraud, and providing false or misleading information.

The need for regulation of the stock markets can be seen by the experiences of the market crash of 1929 and its aftermath, the Great Depression, when there was little federal regulation of securities markets. Many people lost large sums of money, and a run on banks caused many failures.

The Securities Act of 1933 and the Securities Exchange Act of 1934 established SEC oversight and required truthful disclosure by public companies, brokers, dealers, and exchanges. The SEC's Corporation Finance Division also monitors the Financial Accounting Standards Board (SEC 2007).

With the rapid change in financial markets, the SEC is challenged to keep up with new financing mechanisms. For example, it is not sure how to regulate hedge funds that operate across national boundaries.

The SEC also regulates municipal bond trading. This occurs through the Municipal Securities Rulemaking Board, which regulates those who deal in municipal bonds and other municipal securities. This board was established in 1975 by Congress through the Securities Acts Amendments of 1975 to develop rules for underwriting, trading, and selling municipal securities. It is a self-regulated organization subject to oversight by the SEC.

Before the creation of the Municipal Securities Rulemaking Board, the municipal securities industry was largely unregulated. In the 1970s, members of the industry sensed that rapidly expanding bond activity required more formal regulation. Individual investors entered the market in great numbers and there were many new dealers. With the support of the industry and the SEC, Congress passed the Securities Acts Amendments of 1975, which, among other things, created the Municipal Securities Rulemaking Board (Municipal Securities Rulemaking Board 2007).

Development Banks

Around the world, different types of development banks provide funds for infrastructure financing. The World Bank is the most visible example, but many other development banks also operate. Other examples include the Inter-American Development Bank, the Asian Development Bank, and the African Development Bank.

The concept of the development bank is illustrated in Fig. 13-3, where both regular loans and subsidized loans are made to project activities. A regular loan would be repaid at market interest rates, and a subsidized loan would be repaid at less than market rates, perhaps even with no interest. Depending on the degree of subsidy and costs of operations, it is necessary to make up funds from the supporting governments. The bank will also be free to borrow additional funds from the bond market, with these being repaid according to the practices of bond financing.

Applications to Infrastructure and Environment

A knowledge of capital financing, including bond markets and other capital sources, helps the infrastructure and environmental manager assess where to go for investment capital. Capital markets to tap focus on the bond

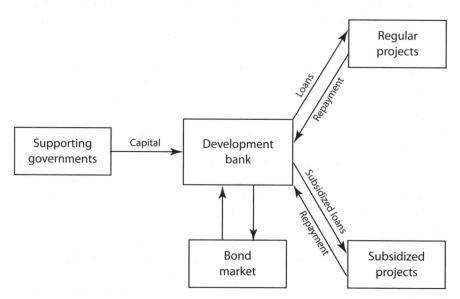

FIGURE 13-3. Development bank operations

markets, notably tax-exempt municipal bonds, which constitute a big part of the total capital picture of U.S. financial markets.

Whereas in the past, municipal bonds were considered staid and reliable investment sources, today they are rated according to risk, and even the most reliable ones are subject to threats from various sources. With the great capital crises of the past decade, strong investment banking firms have folded and private equity sources have been in trouble due to excessive leverage.

A number of factors can affect the interest rate for bond and thus the ability of infrastructure organizations and utilities to service debt. Running these enterprises as businesses is important to make sure that risk is minimized, while at the same time service goals are met.

Development banks can be important sources of capital for some infrastructure and environmental organizations. Though we might think of them as primarily active at the international level and as being mainly intended for developing countries, many versions of them are found at the state-government level as well.

References

Stocks Investing. (2006). A short history of Wall Street. http://www.stocks-investing.com/stock-market-history.html. Accessed July 1, 2009.

Bond Market Association. (2006). http://www.bondmarkets.com/. Accessed August 29, 2007.

Brown, P. (2006). *An introduction to the bond markets.* John Wiley & Sons, New York.

Careers in Finance. (2006). Investment banking: Overview. http://www.careers-in-finance.com/ib.htm. Accessed July 4, 2006.

Goldman Sachs. (2006). Client services. http://www.gs.com/client_services/. Accessed July 4, 2006.

Graham, B. (1996). *Benjamin Graham: The memoirs of the dean of Wall Street,* McGraw-Hill, New York.

Kansas, D. (2005). *The Wall Street Journal complete money and investing guidebook.* Three Rivers Press, New York.

Leigland, J. (1986). "Questions that need answers before we go 'Whoops' again." *Wall Street Journal,* July 10.

Lucchetti, A., MacDonald, A., and Scannell, K. (2006). "NYSE, Euronext set plan to form a market giant." *Wall Street Journal,* June 2.

Maxwell, R. (2007). *Private equity funds: A practical guide for investors.* John Wiley & Sons, New York.

Melicher, R., and Norton, E. (2005). *Finance: Introduction to institutions, investments, and management.* John Wiley & Sons, New York.

Municipal Securities Rulemaking Board. (2007). http://www.msrb.org/msrb1/. Accessed August 29, 2007.

O'Higgins, M. (2000). *Beating the Dow with bonds.* HarperPerennial, New York.

Pristin, T. (2006). "A pot of tax-free bonds for post-9/11 projects is empty." *New York Times*, July 12.

Sakoui, A. (2006). "Eurotunnel discloses debt plan." *Wall Street Journal*, June 1.

Securities and Exchange Commission (SEC). (2007). The investor's advocate. http://www.sec.gov/about/whatwedo.shtml. Accessed March 4, 2007.

U.S. Bureau of the Census. (2007). Statistical abstract. http://www.census.gov/compendia/statab/2007/2007edition.html. Accessed July 25, 2009.

U.S. Internal Revenue Service. (2006). Introduction to tax-exempt bonds. http://www.irs.gov/pub/irs-tege/tebph1a.pdf. Accessed June 17, 2006.

Part III

Economic and Financial Tools for Managers

14

Asset Management for Infrastructure

Economic and Financial Tools

This chapter is the first of four that present tools for infrastructure and environmental management based on economics or finance. This chapter presents asset management, a finance-based organizational tool to manage fixed assets. Chapter 15 presents decision and institutional analysis, a tool set that explains the structure of decisions and how incentives, controls, and other institutional factors influence outcomes. Chapter 16 is about a tool area that is called "engineering economics" but overlaps greatly with the tool area of financial analysis. In conclusion, Chapter 17 sums up how economic and financial tools apply to the common problems of managing infrastructure and the environment, such as finding the balance between public and private approaches.

The Emergence of Asset Management Methods for Infrastructure

The discipline of finance provides us with a comprehensive framework for managing infrastructure over its life cycle. The framework is named after the financial concept of "asset management." For infrastructure systems, this means taking a cradle-to-grave approach to managing fixed assets in order to maximize their productive lifetimes. Thus, the term "asset management," as applied to infrastructure, means managing fixed assets rather than financial resources.

No doubt, a life cycle approach to maintaining and renewing infrastructure can be among the wisest actions that public agencies can take. Older facilities can gain new lives, and scarce funds can be reallocated to more productive uses. With the political necessity to provide essential services even while public investment funds are scarce, asset management for infrastructure offers a powerful tool for managers.

Asset management systems for infrastructure have emerged since about 2000. They are based on a conceptual framework with supporting databases. The conceptual framework is a set of logical and well-known processes, such as maintenance management. The availability of data and information-based management devices has made it feasible to integrate the work processes and thus make asset management systems practical.

In one sense, asset management is not new, because ongoing programs such as capital improvement planning, budgeting, and maintenance management are used in it. However, the integration of these programs with new programs—such as condition assessment, needs assessment, and prioritization—adds value to the approach.

A good answer to the question "What is the main advantage of asset management?" would be that it enables us to make rational management decisions about infrastructure based on realistic calculations of risk. Asset management enables the organization to be data centered through the sharing of information and through processing it to inform the decisions that are made throughout the organization. It helps derive the most value from physical assets, which requires defining and achieving the organization's performance goals. Because most decisions are made along functional lines, integration occurs as you move further up in an organization, and asset management occurs at the level of enterprise management.

This chapter provides explanations to help utilities and other infrastructure organizations understand how asset management works and how it meshes with infrastructure finance. It explains the economic and financial analysis tools that form the core processes of asset management systems.

Definitions of Asset Management

Although competing definitions of asset management have been offered by various agencies, consulting firms, and software developers, these definitions converge on similar concepts. Here is a sample of the definitions:

- ■ "A structured program to optimize the life cycle value of your physical assets" (Harlow and Armstrong 2001).
- ■ "A structured program to minimize the costs of asset ownership while maintaining required service levels and sustaining infrastructure" (Brown and Caldwell 2001).

■ "A combination of tools and procedures to enhance the inventory, management and maintenance . . . of a public works organization" (Carté-Graph Systems 2000).

■ Has the goal to "meet a required level of service in the most cost-effective way through the management of assets to provide for present and future customers" (Champion 2001).

■ A "way of doing business that maximizes the public's return on their investment in utility infrastructure by implementing utility-wide strategies that emphasize reliability in the assets and processes so that the desired levels of service are provided to our customers in the most cost-effective manner" (Seattle Public Utilities definition, quoted by Paralez and Muto 2002).

■ "A business process and a decisionmaking framework that covers an extended time horizon, draws from economics as well as engineering, considers a broad range of assets, . . . incorporates the economic assessment of trade-offs among alternative investment options, and uses this information to help make cost-effective investment decisions" (FHWA 2007).

■ "A structured, integrated series of processes aligned with business goals and values and designed to minimize the life cycle costs and maximize the life cycle benefits of infrastructure asset ownership while providing required performance levels and sustaining the system forward" (EPRI 2007, quoted by Graham et al. 2007).

■ "A strategic and systematic process of operating, maintaining, upgrading, and expanding physical assets effectively throughout their life cycle. It focuses on business and engineering practices for resource allocation and utilization, with the objective of better decisionmaking based upon quality information and well-defined objectives" (Subcommittee on Asset Management, Association of State Highway and Transportation Officials, quoted by Graham et al. 2007).

These definitions show the convergence of concepts of asset management:

■ a life cycle approach,
■ meet required levels of service,
■ cost-effectiveness,
■ short- and long-term strategies,
■ performance monitoring,
■ managing risk,
■ sustainable use of physical resources, and
■ continuous improvement.

The glue that holds these concepts together is data that is organized to support decisions about infrastructure. In other words, an asset management system is a decision-support system with multiple functions.

For a short definition, I offer this one: "Asset management for infrastructure is a method for integrated life cycle facility management across organizations." This definition includes the attributes given above and implicitly includes information-based processes and tools to facilitate efficiency and integration across the enterprise. It requires the organization of data and models to deliver needed decision information to organizational units.

Examples of Infrastructure Systems for Asset Management

Asset management is a generalized concept, so you have to specify what kind of assets are to be managed. Infrastructure assets are mainly physical systems and equipment, and you can also use asset management for financial assets and even natural environmental assets, such as wetlands and aquifers. Our focus here is on infrastructure or what accountants call "fixed assets." As we saw earlier in the book, these assets comprise the major share of the capital base of utilities or public works agencies because infrastructure systems are capital intensive.

The categories of infrastructure identified in Chapter 1 offer a classification system that can be used to explain different approaches to asset management. As examples:

- *Highways and bridges:* The U.S. Federal Highway Administration (FHWA 2008) has an Office of Asset Management, which promulgates advice and works with state departments of transportation to implement the systems.
- *Electric power systems:* The Electric Power Research Institute (2007) has taken the lead to create aids and instructional material for utilities.
- *Water and wastewater systems:* The U.S. Environmental Protection Agency (2008) promulgates guidance. Also, the Water Research Foundation and the Water Environment Research Foundation (2008) are active in developing asset management technologies. They worked to create a software package titled "Sustainable Infrastructure Management Program Learning Environment" (SIMPLE) to advance asset management and are continuing to develop new products.
- *Facilities asset management:* Given the large capital investment in facilities of all kinds, asset management tools are available from a number of organizations involved with facilities management.

A Model for Asset Management

On the basis of a review of models for asset management, Graham and colleagues (2007) identified common practices used across the platforms of different infrastructures. By adding management functions such as capital budgeting and accounting, a general model for asset management can be created (Fig. 14-1). This model provides the comprehensive framework needed to organize an asset management program.

Organizational functions included in the model are for planning, finance, engineering, and operations. The framework provides the mechanism to integrate them on the basis of shared use of data.

Decisionmaking is centered in the financial function of the model because asset management decisions ultimately involve finances. The finance staff are stewards of financial information, but decisions are made by managers and policymakers who use the outputs of the asset management information and analysis.

Planning Tasks

Planning is an integrative activity itself and involves assessing needs, preparing plans, and monitoring performance of systems. Part of planning is the prediction of future needs, which requires condition assessment and other

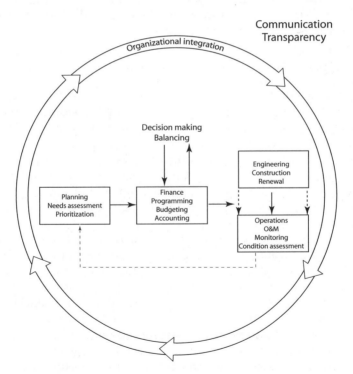

FIGURE 14-1. Conceptual view of a model of enterprise asset management

planning data. At the national level, examples can be found by federal agency programs to assess needs for drinking water, water quality systems, highways, and transit, all of which have established processes for needs assessment.

Systematic, rational, and comprehensive priority setting is based on an examination of future needs and balancing risk by addressing trade-offs. If you can quantify risk, then you can estimate the cost to mitigate it. Priority setting is where economic and financial tools are used directly as you look for plans that yield maximum benefits and rates of return (see Chapter 16).

Needs assessments consider goals and requirements as well as the ability of assets to meet their demands. Gap analyses are used to identify the differences between what should be and what exists. The asset management system should yield facility information for needs assessments linked to the stages of planning, from master planning to detailed facilities planning.

Examples of the use of economics and finance for planning include priority setting for local street improvements and schemes to replace water pipes before they break. In each case, the condition of facilities is assessed. On the basis of analytical studies of performance variables, you can determine the life cycle cost of management decisions to improve facilities or to do nothing. This leads to needs assessments, which can be arrayed according to priority. Investment schedules are then set on the basis of the management organization's goals and resources.

Financial Tasks

The financial functions involved in asset management center on budgeting, accounting, and performance assessment. The capital budget process, with its planning-programming-budget features, can be used to organize financial controls related to fixed assets.

The planning process feeds information about needs and priorities to the budget process, which becomes the decision point for deciding if infrastructure needs can be financed through the capital budget or not. In some cases, lower-cost needs such as minor maintenance can be met in the operations budget.

Theoretically, under the Government Accounting Standards Board's "GASB 34" rules, accounts should reflect the condition of the fixed assets so that the financial records align with the records of the engineering planners, but this level of consistency remains an elusive goal. This was discussed in Chapter 10. Regardless, the manager's goal should be that financial records become more reflective of the true costs of deferred maintenance.

The budget and financial reporting processes offer managers places to record the results of decisions about fixed asset. In budgeting, the results of performance assessments can be noted to improve the allocation of funds in future cycles. For example, if the data show that not enough funding was provided for water main renewal and that excessive main breaks occurred, then the budget can show the need for more funds to meet this need. If the data show that road mobility has or has not been improved, the comprehensive annual financial report can publish the performance information for decisionmakers to consider.

Engineering Tasks

Engineering is a core participant in asset management and is most likely the custodian of data on asset inventory, condition, and performance. Its basic functions are the design, construction, and inspection of physical assets. These provide asset management information through studies, surveys, maps and the geo-database, and related construction documents.

These information products can be made available electronically in the asset management system. As-built drawings from engineering should be accessible online to maintenance forces, and planners should be able to pull up older studies and investigations for needs estimates. This enables the asset management system to integrate the planning-engineering-maintenance cycles of infrastructure, as well as the planning-financing-renewing cycles.

Operations and Maintenance Tasks

The operations function handles monitoring and much of condition assessment. It works closely with engineering to provide data and act upon recommendations based on asset studies. Maintenance management is closely associated with operations, with the goal of ensuring maximum performance and obtaining the highest yield from assets. Maintenance management relies on inventory and condition assessment functions. Maintenance forces provide data to the information management function, which then provides it to other sections.

Condition Assessment for Infrastructure and the Environment

The process of "condition assessment" is at the heart of asset management. Assessing the condition of any facility or system is essential so that gaps between the condition and the standard are measured and investments can be planned.

Condition assessment is the key to effective risk management and to being proactive in preventing the failure of systems. For infrastructure facilities, the use of condition curves can help communicate the process of asset deterioration and the need for renewal. The condition curve for a facility is like a depreciation curve in accounting. In accounting, you can have straight-line or accelerated depreciation curves. These would compare with the standard condition curve, was as shown in Fig. 4-7.

For infrastructure facilities, the evidence shows that once the break-in period is past, the condition deteriorates with time in a regular pattern. This is evident from pavement condition curves, which look like the basic curve in Fig. 4-7 and which are often used to explain normal declines in condition.

Another common shape of condition curves is the "bathtub" functional shape (Fig. 14-2), which explains the behavior of water main breaks. It graphs and labels the phenomena of a break-in period, followed by normal service life, followed by more frequent failures.

Condition curves can also be used to illustrate a facility's environmental condition, which starts with a pristine or natural state. Degradation occurs slowly through use, and it will continue until you reach ruined conditions unless a limit is imposed by regulatory controls. The gap between the pristine condition and the sustainable level is the renewable capacity that can be exploited, and the level remaining is that required to sustain natural systems (Fig. 14-3).

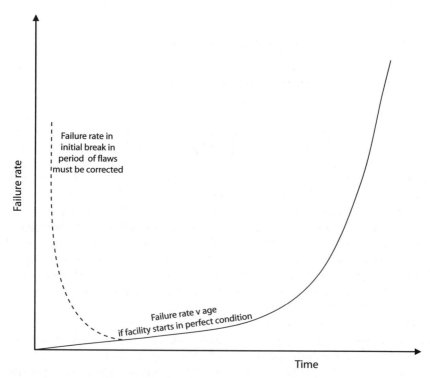

FIGURE 14-2. Correlation of condition and bathtub failure condition curves

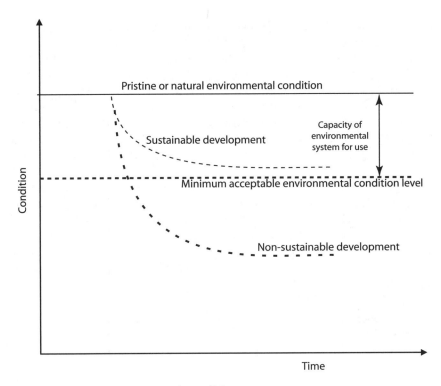

FIGURE 14-3. Environmental condition curves

Integration in Asset Management

Integration is the key attribute that separates asset management from its component tasks, such as maintenance management. In the asset management model, several organizational drivers help implement integration.

In a general sense, coordination is the starting point. It means to achieve harmony among diverse interests, and it is an integration mechanism at whatever level it occurs. Levels of coordination can vary from simple discussions among workers to elaborate computer-based reporting systems.

Three elements are primary drivers in integration: the establishment of integrated goals and strategies, a common data platform, and a transparent assessment method. To achieve these, commitment by management to coordinate decisions is required, and this in turn requires good communication, meetings, shared values, and other similar practices.

Asset management integrates the financial aspects of life cycle management, as shown in Fig. 14-4 for infrastructure. It achieves this by managing

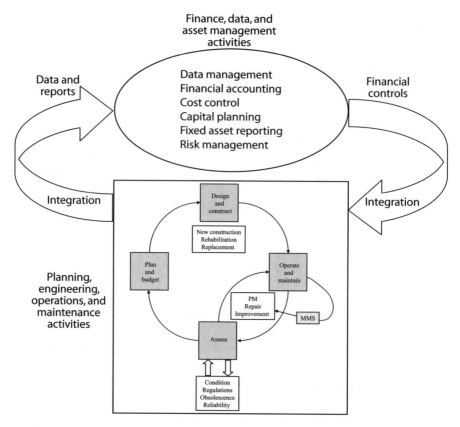

FIGURE 14-4. Infrastructure life cycle and asset management

capital and operating costs over infrastructure life cycles to achieve maximum yields from assets during their lives.

In asset management, planning, engineering, operations, and maintenance are linked to finance and to each other through the sharing of data.

These elements of integration are illustrated in Fig. 14-5, which shows the integrative drivers as ellipses. They are integrated goals and strategies driven by the planning function, a common data platform for the enterprise, and a valid program to assess results. Further integrative activities by management are commitment and shared values, coordinated decisions, and communication and transparency.

A common data platform is the most important feature of asset management systems. Without data integration, asset management lacks any punch. This helps in establishing a common understanding through shared

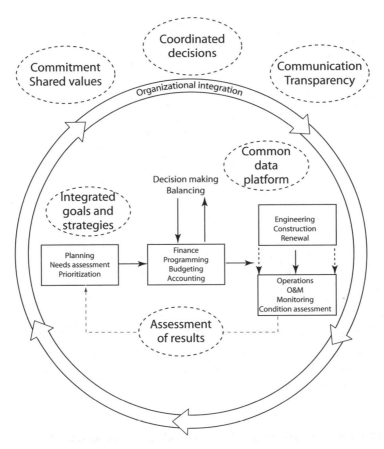

FIGURE 14-5. Asset management model showing integrative drivers

data and information. The use of commercial software packages to integrate data is becoming more common. They commonly include many types of management data, such as work order management, along with infrastructure information.

The relationships between organizational form, functions, and the use of management information continue to evolve. Before computers, organizations were fragmented because information could only be shared among a few people at a time. Organizations had many layers, so information could only be passed smoothly to small sections of workers. Now, however, information can pass over networks and be more widely available. This has flattened organizations and eliminated middle management positions, whose main functions were to process information.

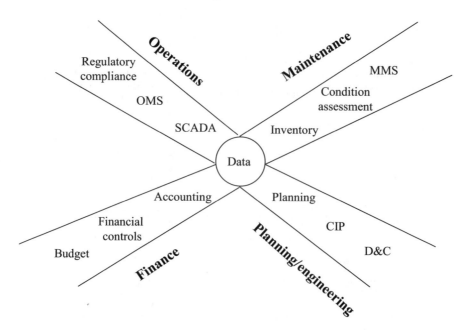

FIGURE 14-6. A data-centered infrastructure management system

Note: OMS is operations management system, SCADA is supervisory control and data acquisition, MMS is maintenance management system, CIP is capital improvement program, and D&C is design and construction.

Computers enable workers to share information among functional departments in organizations. A data-centered infrastructure management system (Fig. 14-6) will link functional areas of the organization to work on planning, engineering and construction, budgeting and finance, operations and maintenance, and information systems. This enables departments to work together on shared, cross-cutting objectives, such as asset management.

In a practical sense, sharing information is harder than it seems. Because obtaining enterprise-wide databases and geographic information systems requires managers and workers to share information and authority, it might generate resistance. Asset management offers great possibilities to improve effectiveness, but the real-world aspects of organizations and fragmented data sources pose significant challenges.

Asset management has become an international tool for managing infrastructure and other fixed assets. Some of the more interesting innovations have come from Australia, which, for example, has produced a comprehensive *International Infrastructure Management Manual* (Institute of Public Works Engineering Australia 2006).

Applications to Infrastructure and the Environment

Infrastructure managers can use asset management tools to extend the lives and improve the performance of their constructed facilities. They can use the model for asset management that was presented as a structure for their software and management protocols.

This chapter has presented a clear definition of asset management to enable managers to see its elements. Examples of asset management systems for categories of infrastructure illustrated how to apply the tools in different situations and with various platforms.

The chapter illustrated how asset management tools can be used for planning, engineering, finance, and operations and maintenance tasks. It also showed how condition assessment technologies are essential to implement effective asset management systems. Above all, it illustrated why data-centered integration represents the greatest reason for added value through asset management tools.

References

Brown and Caldwell. (2001). *Quarterly* (Walnut Creek, CA), Summer.

CartéGraph Systems. (2000). *Getting started in public works asset management.* CartéGraph Systems, Dubuque, IA.

Champion, C. (2001). "Asset managers or merely asset owners?" *APWA Reporter*, May, 13.

Electric Power Research Institute. (2007). *Information technology for enterprise asset management: An assessment guide.* Electric Power Research Institute, Palo Alto, CA.

U.S. Federal Highway Administration (FHWA). (2007). Asset Management. http://www.fhwa.dot.gov/infrastructure/asstmgmt/assetman.htm. Accessed July 10, 2007.

———. (2008). Office of Asset Management. http://www.fhwa.dot.gov/infrastructure/asstmgmt/. Accessed May 24, 2008.

Graham, A., et al. (2007). Review of asset management practices and needs in the North American water and wastewater industries: A status report. Paper presented at American Water Works Association Distribution Research Protection Symposium, Reno, March 2–3.

Harlow, K., and Armstrong, B. (2001). "What's in the pipe for collection system asset management? GASB 34 and CMOM—Rumbles." *Rocky Mountain Section of AWWA and RMWEA*, 41(2), 4–8.

Institute of Public Works Engineering Australia. (2006). International infrastructure management manual. http://www.ipwea.org.au. Accessed July 9, 2007.

Paralez, L. L., and Muto, D. (2002). "An asset management template." In *Assessing the future: Water utility infrastructure management*, ed. David M. Hughes. American Water Works Association, Denver, CO.

U.S. Environmental Protection Agency. (2008). About asset management. http://www.epa.gov/OW-OWM.html/assetmanage/. Accessed May 24, 2008.

Water Environment Research Foundation. (2008). *A holistic approach to asset management using SIMPLE knowledge base for water and wastewater operations.* Web Seminar Series. Water Environment Research Foundation, Alexandria, VA.

15

Decision Analysis and Institutional Analysis

Decisionmaking for Infrastructure and Environmental Systems

Chapter 1 explained the importance of effective decisions about infrastructure and the environment and that the decisions require infrastructure and environmental managers to use tools from economics and finance to understand consequences and evaluate alternative strategies. The decisions are made in arenas that also include politics and human behavior, and it is important to understand the contexts in which the economic and financial tools can be used.

This chapter explains these contextual situations within the frameworks of decision analysis and institutional analysis. Decision analysis introduces relevant questions into the equation—such as What is this decision? Why is it important? How is it made? Who makes it? and When is it made? Institutional analysis considers questions such as How reliable is the decision information? What happens if the assumptions change? How strong is the management authority? and Which incentives drive the behavior of the participants?

These questions and elements of analysis are part of the disciplines of economics and finance. In economics, the fields of public choice theory and political economy address these issues. In finance, the process of rating bonds for risk takes into account institutional factors. Also, the Government Finance Research Center (GFRC) advises researchers to consider institutional questions as part of an overall financial analysis (GFRC 1981). This chapter illustrates how to apply analysis and objective thinking to public decisionmaking by applying them in considering all the questions, not just those that can be quantified.

This is urgent when financial risk is present, because a failure to meet expectations can be measured by the numbers and there is often little opportunity for corrective action until too late. This is why institutional analysis is so important. A way to think about this is to understand ongoing problem analysis as to some extent being linear—that is, as considering events in a logical and sequential manner. However, the world often does not work this way but operates in a nonlinear and unpredictable manner. Institutional analysis is an attempt to figure out how the world really works. Figure 15-1 illustrates the quandary of a problem solver who would like to follow a systematic process but is confronted with perplexing institutional problems.

Elements of Decision Analysis

Infrastructure decisions involve elements that go beyond pure numbers and require tools such as multiobjective decision analysis and institutional analysis. For example, to explain how to plan for wastewater facilities, the GFRC (1981) listed the necessary tasks as revenue analysis, cost analysis,

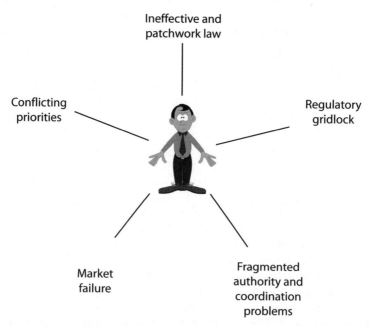

FIGURE 15-1. The quandary of a problem solver confronted with perplexing institutional problems

institutional analysis, ability-to-pay analysis, secondary impacts analysis, and sensitivity analysis. Of these, all except revenue and cost analysis involve aspects of decisionmaking that go beyond the numbers themselves.

Revenue analysis identifies the revenue sources available and their feasibility, on both a financial and a political basis. Cost analysis considers construction costs, operating and maintenance costs, and other costs such as regulatory rules and planning. Cost analysis is used to cut waste from a system and to determine the components of cost to assign to different users. Direct costs—such as wages, equipment, operations and maintenance expenses, depreciation, and capital recovery—can be assigned to services. Indirect costs, such as administrative support services, cannot be attributed to specific services, but they can be allocated over the services they support (American Water Works Association 1995).

Once an issue is understood at a basic level, a problem-solving process can be initiated by following steps that answer these questions:

■ What is the problem?
■ What are the goals for solving it?
■ What are the measures of success in solving it?
■ What are the alternative solutions?
■ What are the ratings of the alternative solutions compared with the goals?
■ What is the decision?
■ How will it be implemented?

To answer these questions requires both "hard" and "soft" analysis. Hard analysis follows a mostly linear process that utilizes the tools of management science. Soft analysis considers "what if" questions and political/institutional elements. It includes areas such as human resources, regulatory surprises, unforeseen events, and other issues, which effectively determine the ability to pay, secondary impacts, sensitivities, and other institutional issues.

Looking at the decisionmaking process shows that decisions have more subtle aspects than meets the eye. In particular, the first three questions listed above may require multiple iterations to get them right. These questions can be restated this way:

■ What *really* is the problem?
■ Are the goals for solving the problem clear, or must they be discovered?
■ What are the measures of success in solving the problem, *and how will we know when we have succeeded*?

Answering these questions requires more than just linear thinking, and the process of institutional analysis provides a systems view that enables analysts to look at the whole picture and answer "what if" questions.

Institutional Analysis

Institutional Issues

Looking at the tasks identified by the GFRC, we can see that other than revenues and costs, the factors deal with organizational stability, ability-to-pay analysis, secondary impacts, sensitivity analysis, and other institutions that might affect the outcome of the venture.

The "institutional analysis" called for by the GFRC focuses on the management organization for wastewater service, which is only one of the institutional questions to consider. The term is derived from the verb "to institute," and an institution is something that has been established. The term generally means the rules and structure of the game, and it includes organizations (such as a transportation department), buildings and facilities (such as a mental health facility), customs and practices (such as an unwritten rule), relationships (such as a "gentleman's agreement" to share water), and laws (such as a minimum wage law). In the case of infrastructure and the environment, institutions usually relate to government in one way or another.

Ability to pay, secondary impacts, and sensitivity analysis are related to the institutional frameworks within which decisions are made. Ability to pay considers the capability of the management organization to bear the cost of the service, as in the use of a financial ratio to show the limitation on debt of a local government as a percentage of assessed valuation. Secondary impacts consider economic, social, and environmental factors. Sensitivity analysis examines changes in the outcomes of the analysis that result from changes in the assumptions.

Institutional Analysis for Infrastructure and Environmental Decisions

The particular issue addressed by the GFRC was whether to approve a grant or loan to an organization building a wastewater treatment plant. This is much like the issue facing a commercial banker who is trying to decide whether a borrower can repay a loan and is a matter of risk. With infrastructure and environmental decisions, many of the issues are in the public sector and, though they involve risk, it is not as well defined as the banker's decision.

The decision of the banker or the agency granting money for the construction of a wastewater treatment plant is based on the lender's or grantor's financial role as steward of someone else's money. In a broader sense, any time an agency makes a decision about infrastructure and/or the environment that involves the allocation of public resources, it should consider risk in the same way. That is, the allocation of water, environmental quality, or access to natural resources means allocating the public's capital, much like public investments in infrastructure facilities.

Institutional analysis to determine the ability of an organization to repay a loan or utilize a grant is one example of analysis that can be applied to solve the institutional problems of infrastructure and the environment in a more general way.

Key Questions for Institutional Analysis

Decision analysis that incorporates institutional questions is complex, but on an overall basis it can be reduced to five key questions: What? Why? How? Who? When? These questions are listed here with sample follow-up questions in italics that focus on from the issue of whether to approve the wastewater grant:

- What is the issue? *Should the grant to build the wastewater treatment plant be approved?*
- Why is it important? *Which public interest questions are involved?*
- How does it work? *How will the money flow?*
- Who has what role? *What are the key roles and responsibilities? Have the parties agreed to fulfill their roles?*
- When are actions required? *How must the process work?*

These questions should be addressed in the context of a problem-solving process. The first question, what is the issue, is directed toward identifying the problem to be solved. The last question, what actions are required, requires analysis itself to identify the required actions. In between are several steps that require definite analysis to avoid an unstable process that might occur unless the problem and strategies are clearly understood.

A Framework for Institutional Analysis

At its root, institutional analysis is concerned with human behavior and institutions. Ziegler (1994) described its behavioral dimension as the "correlated patterns of human activity in groups," the "rules of the game," and means of modifying behavior by altering the patterns that direct it. These are very general notions, and we can speak specifically of the institution of concern, for example, a management agency for infrastructure, the regulation, or the group that exercises control.

A framework for institutional analysis must constitute a systematic and repeatable way to break the elements apart so they can be studied. For Ziegler's (1994) system of analysis, he presented a list of key questions, which are stated here in simplified form:

- What goes on here?
- What processes need adjustment?

■ What problem-solving know-how is available?
■ What ought to go on here?
■ What are the impacts of change on other patterns of activity in this institution?
■ What are the impacts of change on other institutions?

These generic questions provide the basis for a "gap analysis," which is needed for infrastructure needs assessments.

Another angle on the analysis method is to ask questions in each category of institutional element. This list illustrates questions in the range of institutional arrangements that are usually encountered in the infrastructure and environmental arenas:

■ What are the laws and controls—what is the legal framework and what are the control mechanisms (laws, regulations, decision requirements, enforcement mechanisms)?
■ Who has control—who are the designated authorities and stakeholders (mainly organizations)?
■ What are the incentives—such as ownership, property rights, and incentives?
■ Who has what role—what are the roles, responsibilities, and relationships between stakeholders?
■ What is the management culture—what are the management practices, customs, and ways of doing business (informal institutions)?

Taking these two lists together gives us the framework for institutional analysis. For the infrastructure system being studied, this framework includes:

■ a conceptual model of how the management and control system work (What goes on here?);
■ identification of the key issues in each category of institutional element (What processes need adjustment?); and
■ identification of institutional practices that should lead to improvement (What ought to go on here?).

Gap Analysis

A gap analysis (Fig. 15-2) is a comparison of an existing situation with a desired situation. It is regularly used in infrastructure and environmental studies to identify problems that require solutions. Actually, any study about management or organization that leads to recommendations for improvement is a gap analysis. A common example is a needs analysis, which often precedes budget exercises. Gap analyses are used in public administration,

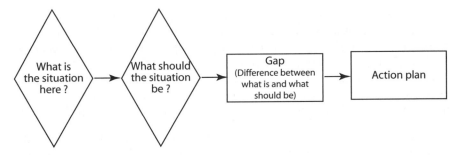

FIGURE 15-2. Gap analysis and response

which focuses on policy analysis and on institutional improvement. As a field, public administration is closely allied to civil engineering, because both focus on government actions. Policy analysis is a systems approach to decisionmaking, with an action orientation (Kraemer 1973).

The institutional aspects of infrastructure situations refer to gaps in policies, responsibilities, problem identification, decisions, and actions that are required to maintain adequate service levels. Certain patterns of these areas lead to good service, and others do not. The differences between the two patterns constitute institutional "gaps." The patterns leading to success might be called "best management practices," another widely used but ambiguous term.

Applications to Infrastructure and the Environment

The central work of managers is making decisions, and it can be difficult to break through the cloud of confusions to see the real elements of a problem. This chapter helps in decision analysis by showing how a series of questions frame problems and can lead us toward the best solutions. It is a practical chapter, and it enables us to formulate these key questions as this series:

- What is the problem?
- What are the goals for solving it?
- What are the measures of success in solving it?
- What are the alternative solutions?
- What are the ratings of the alternative solutions compared with the goals?
- What is the decision?
- How will it be implemented?

The chapter goes on to probe these questions further to find out what the problem *really* is, whether the goals in solving it are clear or must still be discovered, and how we can know when we have succeeded.

The techniques examined in the chapter are named decision analysis and institutional analysis to illustrate that they deal with questions of incentives, structure, authority, and barriers. Some people call these questions "politics," and if anything, they represent the nontechnical aspects of the questions. These techniques are useful in planning because they can help us to create analyses of gaps that can be used to argue for approvals, investments, and other decisions by the authorities in particular situations.

References

American Water Works Association. (1995). *Water utility accounting.* 3rd ed. American Water Works Association, Denver, CO.

Government Finance Research Center (GFRC). (1981). *Financial management assistance program: Planning for clean water programs—The role of financial analysis.* U.S. Government Printing Office, Washington, DC.

Kraemer, K. L. (1973). *Policy analysis in local government.* International City/ County Management Association, Washington, DC.

Ziegler, J. A. (1994). *Experimentalism and institutional change: An approach to the study and improvement of institutions.* University Press of America, Lanham, MD.

16

Engineering Economics and Financial Analysis

Economic and Financial Analysis Tools

Economics and finance have a common core set of quantitative analysis tools that focus on using compound interest to compare the rates of return of alternative strategies. Many of the tools can fit within both fields, so this chapter is titled "Engineering economics *and* financial analysis." The separate names of engineering economics and financial analysis developed because engineers were pioneers in analyzing alternative investments, and the field of finance developed later.

Engineering economics is as much about finance as it is about engineering. Though engineers take courses in engineering economics and solve problems about it on their licensing exams, other people use the same methods to solve problems ranging from personal finance to business transactions. A financial calculator even looks like a science and engineering calculator, but it has different built-in functions.

The common analysis techniques of engineering economics and financial analysis are based on computing cash flows over time. Engineering economics focuses on engineering problems and can be extended to social and environmental objectives, whereas financial analysis usually focuses on narrower questions of financial inputs and outputs.

Computing rates of return for infrastructure and environmental projects also involves social and environmental perspectives. These require multiobjective analysis to consider environmental and social issues, as well as financial issues. For a prime example, a multicriteria decision analysis table, with social, economic, and environmental accounts—which is explained below as an analysis tool—is a way to distinguish the broader concept of economic analysis and the narrower concept of financial analysis. Financial analysis deals with the economic accounts that consider actual money flows. Other economic accounts might deal with economic development, equity, and a range economic benefits that do not involve money flows.

The Evolution of Engineering Economics and Financial Analysis

The common roots of engineering economics and financial analysis date back to the nineteenth-century analysis of railroad projects, which consumed large amounts of capital. According to Grant and others (1997), the first book on engineering economics was an 1887 work titled *The Economic Theory of Railway Location* (Wellington 1887). At that time, bankers and financiers had been using financial analysis for a long time, but the academic field of finance was in its infancy.

During the twentieth century, the fields of engineering economics and finance evolved in parallel through professors, books, courses, and the fields' own learning channels. The field of industrial engineering embraced engineering economics as a central issue. As business education developed, it also embraced financial tools. Today, engineering and business education are broadening, each carrying its own economic and finance tools with it.

Engineering economics can merge with finance and management disciplines. For example, at Stanford University (2006), the Engineering Economics Department merged with the Management Science and Engineering Department, which includes eight related areas: decision analysis and risk analysis; economics and finance; information science and technology; optimization and tools of system analysis; organizations, technology, and entrepreneurship; probability and stochastic systems; production and operations management; and strategy and policy. These topics are as close to business education as to engineering.

Financial analysis has also evolved with the field of accounting, which was explained earlier in the book. Finance was never simple, but information technology, trends toward complex accounting rules, new financial instruments such as derivatives, and an international monetary system have made the field of finance much more complex.

From Analysis to Decision Support

The question that is usually posed to engineering economists or financial analysts is "What makes the best investment decision?" For example, should we build a new road now or repair the old one again? Should we increase the size of the flood control system, or should we take a risk on a flood occurring? Should we build, buy, or rent our building?

The starting tool to apply for these questions is decision analysis, which means a systematic way for evaluating the pros and cons of different decisions. When applied to public sector decisions, it must consider economic, social, and environmental goals as well as financial rates of return. Chapter 15 explained how decision analysis must also consider institutional factors.

Multicriteria decision analysis can be used as a framework to organize evaluation information for different goals. It can incorporate economic criteria, social impact analysis, and environmental impact analysis to evaluate nonfinancial impacts. It can display multiple objectives on a single decision document and thus facilitate the decisionmaking process.

Decision analysis is organized through the planning process. Its basic steps are the identification of the problem and evaluation criteria, formulation of alternatives, evaluation of alternatives, decision, and implementation. Though these are the basic steps, decision analysis occurs in a political framework, which wraps the basic steps of planning in an environment where stakeholders compete to advance their agendas.

In addition to multicriteria decision analysis, the steps in planning require quantitative techniques such as cash flow analysis, benefit-cost analysis (BCA), and the discounting of money to discern the advantages and disadvantages of various courses of action involving capital investments, whether in the private sector or public sector. In particular, engineering economics and financial analysis are used in the evaluation stage of the planning process.

In financing a public infrastructure system, the capital requirements might be for a new road or water treatment plant. In the private sector, it might be investment for a new manufacturing plant or equipment. The best investment in the private sector involves finding the highest rate of return, usually to improve the bottom line of a business venture. For the public sector, investing public capital to obtain the highest return requires taking into account societal objectives that may or may not be quantifiable with interest rates.

Multicriteria Decision Analysis

Multicriteria decision analysis (MCDA) provides a framework for organizing the evaluation information for different categories of goals to consider trade-offs among them. It evolved from economic, environmental, and social impact analysis, as expressed through welfare economics and utility theory (see Chapter 2).

Chapter 2 explained how welfare economics studies the maximization of public or social welfare by seeking the best value of a social welfare function, which includes categories of public goods such as economic development, environmental quality, and improved quality of life. It also explained how utility theory seeks similar goals where a person's utility measures satisfaction from some outcome and decisions are made by choices that increase a person's or organization's utility.

Hill (1968) was an early developer of an MCDA for an infrastructure decision problem. He showed how a "goals achievement matrix" could be used to display the benefits and costs of a transportation problem. At the time he presented his work, it was actually a Ph.D. dissertation in the urban planning school at the University of Pennsylvania.

MCDA measures how different strategies or projects lead to achievement in different categories of goals. Though this seems straightforward, it is not simple to reduce information in this way. You end up with much numeric data that seeks to measure outcomes and preferences, but people have different opinions about what the numbers should be. For this reason, at the end of the day, an MCDA exercise is usually considered advisory, and the decisionmakers vote or debate to bring out sensitivities in the assumptions.

In its simplest form, an MCDA display shows how strategies or projects score in the goal categories, as shown in the sample matrix given as Fig. 16-1. In the matrix, you provide a net score or a descriptive analysis for each project in each category. To do this, you must be able to evaluate the projects to determine the scores, and you must have a scoring system.

Economic evaluation uses BCA tools. Environmental evaluation uses environmental impact analysis, and social evaluation uses social impact analysis. Benefits and costs can be quantified in dollars, even though the estimates are often uncertain and inexact. Environmental and social analysis outcomes are harder to quantify and often rely on verbiage rather than numerical scores to describe positive and negative impacts.

Using Benefit-Cost Analysis to Evaluate Economic Feasibility

BCA is used to compare the potential economic and financial merits and demerits of alternative projects. As an analysis technique, it is known to have emerged with the 1936 Navigation or Flood Control Act, which required that

	Economic	Environmental	Social
Project A			
Project B			
Project C			

FIGURE 16-1. Matrix for scoring a multicriteria decision analysis

projects could be authorized only if their benefits exceeded costs, regardless of to whom they accrued. This concept from the New Deal era sought to provide a mechanism for public investment to benefit the nation at a time of economic hardship. It was up to the Army Corps of Engineers to figure out how to implement it. To be fair, BCA's concepts had evolved from earlier economic thinking, and some give credit to Jules Dupuit, a French engineer and economist (Hager 2004).

The concepts of financial costs and benefits are easy to grasp when you explain them as income and outgo or revenues and expenses. This takes care of the financial side, even when it is called engineering economics. On the economic side, the concepts are harder to grasp. In the economic version of BCA, a benefit is a gain and a cost is a loss. You have to specify what kind of gains or losses, how large they are, who they impact, and when they occur.

Benefits and costs may be either tangible or intangible and direct or indirect. Tangible benefits are those you can measure, like an increase in profit from an investment. An intangible benefit cannot be measured directly. An example would be an increased sense of security because a dam was strengthened and is less likely to fail. A direct benefit is one that stems expressly from the purpose of a project, such as reduced travel time because of a road improvement. An indirect one stems from other, perhaps unintended, purposes, such as the economic benefits created by that road for the owners of nearby properties.

Table 16-1 illustrates a few examples of tangible and intangible costs and benefits in the categories of infrastructure. How large the benefits and costs are require studies and evaluation to quantity them. In many cases, this can involve expert opinion and still remain controversial because of uncertainties, valuation methods, and different assumptions of the experts.

TABLE 16-1. Sample of benefits and costs for different infrastructure categories

Category	Example of benefit	Example of cost
Built environment	Provide amenities to public	Loss of historical building
Water resources	Stop flood losses	Financial project cost
Transportation	Save driving time	Add to noise level
Energy	Obtain lower-cost supplies	Add to air emissions
Waste management	Improve water quality	Degrade water quality
Environment	Improve habitat	Lose habitat

Who the benefits and/or costs fall on is called their "incidence." This is important because if the beneficiaries pay the costs of a project, that is one thing, but if society as a whole pays them, regardless of who benefits, that is really a redistribution of wealth.

The times at which the benefits and costs occur are taken into account in discounting with interest rate factors. When the benefits and costs are monetary, this is relatively simple, if you know the discount factor. However, often the benefits' and costs' times of occurrence are uncertain.

To compare the benefits and costs (gains and losses) from proposed actions, you must value them. This requires some kind of measure of relative merit, hopefully dollars.

Benefits are measured by dollars produced in categories of benefits, and costs are also measured in dollars. Once benefits and costs are reduced to the same time basis and made commensurate, they are usually compared on the basis of the benefit-cost ratio or net benefits. Categories of benefits in a water project might include, for example, the municipal and industrial water supply, agriculture, urban flood damage, hydroelectric power, navigation, recreation, and commercial fishing. In a transportation project, they might include travel time savings, energy savings, and reduced maintenance.

Below, I discuss how to discount money to account for interest rates over time. BCA is sensitive to discounting techniques, and selecting the proper interest or discount rate can introduce controversy into analysis. Whereas in financial analysis, the interest rate is the cost of money, economic analysis requires that social purposes also be considered. This requires the use of a "social discount rate," which is difficult to determine. Guidelines for social discount rates to evaluate federal programs are set by the U.S. Office of Management and Budget, but for water projects they are in a separate publication, *Economic and Environmental Principles and Guidelines for Water and Related Land Resources Implementation* (U.S. Water Resources Council 1983). Though decisionmakers do not always accept results of BCA because of difficulties in estimating benefits and costs, if consistent techniques are used, projects with greater merit show up better on a relative basis.

Environmental and Social Analysis

In concept, the principles of economic analysis can apply to environmental and social analysis as well as to matters of money. However, financial discounting using interest rates is more difficult to apply to environmental and social issues than to financial situations. To see this, consider that the interest on money is a measure of its value in use. If a lender provides $100 in capital to a borrower, the loan might earn $3 in annual interest. To pay the interest, the borrower must apply the $100 to a beneficial purpose to earn at least the $3. You also have environmental and social "capital," and this capital will also earn returns, but methods to analyze these returns are not well developed. For this reason, rather than analyze returns on environmental and social capital resources, we analyze impacts on them from actions.

Environmental Impact Analysis

Environmental impact analysis (EIA) is a way to compare projects on the basis of their environmental impacts. Though its basic concepts, such as measuring if a project reduces the wildlife habitat, might seem intuitive, EIA is a complex process, and national guidelines have been developed under the 1970 National Environmental Policy Act (NEPA), which established goals for environmental policy and requirements for environmental impact statements (EISs) for major federal actions that affect the environment. It also established the Council on Environmental Quality to review policies and programs and to prepare the president's annual environmental report to Congress.

An EIS evaluates the environmental impacts of a proposed action, including its unavoidable adverse environmental effects and alternatives to it that are available. The president's annual environment report describes the condition of the nation's air, aquatic, and terrestrial environments. Since NEPA was passed in 1970, the EIS process has influenced many projects and actions. On the positive side, it provides for the coordination of the inputs of diverse interests and thus improves planning. On the negative side, it can be bureaucratic, expensive, and time consuming.

Social Impact Analysis

Social impact analysis (SIA) is another way to compare projects. It is not used as much as economic and environmental analysis tools, and its techniques are not as standardized. Basically, SIA would measure how a strategy or project would affect people. To assess that, you would need to identify the stakeholder groups to be impacted, what the impacts might be, and how the projects would affect the groups in the categories of impact, along with making a comprehensive analysis of the results.

Principles of Engineering Economics and Financial Analysis

Chapter 11 explained financial planning and analysis as they occur in the public finance field. Of the six steps in financial planning, each has "analysis" as part of its title—revenue analysis, cost analysis, institutional analysis, ability-to-pay analysis, secondary impacts analysis, and sensitivity analysis.

I also discussed "decision analysis," which means about the same thing. "To analyze" means to break something into its parts to study them. In this sense, both engineering economics and financial analysis involve breaking a possible deal into its parts to consider them.

As an academic subject, engineering economics has been taught to engineers for many years. Generally speaking, it entails applying economic criteria to the decision analysis of engineering problems. Its major innovations in recent years have been spreadsheets and computers, which make computation easier.

This section presents an overview of engineering economics and financial analysis and their main computational procedures. The presentation is not exhaustive, and many detailed texts are available on the subject, such as Grant and others (1997) and Newnan and others (2004).

Grant and others (1997) presented a set of principles for formulating problems. These seem intuitive and include such guidelines as: Decide among clear alternatives, and consider the merits of all appropriate alternatives; decide on the basis of expected future consequences; clarify the viewpoint for weighing the merits of alternatives; weigh only commensurable money consequences on the same time basis; consider only differences among alternatives; make separable decisions separately; have criteria for decisionmaking, so as to make best use of resources; anticipate uncertainty and use secondary criteria to aid making decisions about the future; consider the nonmonetary consequences as well as the financial consequences; and consider the side effects. Taken together, these principles can be expanded to create a guidebook for financial analysis and are very useful for decision analysis.

For engineers seeking to apply MCDA to decisionmaking, considering the nonmonetary consequences as well as the financial consequences and side effects of a possible action comes naturally. In fact, environmental and social factors are often intangible; that is, they cannot be measured in dollars, so considering them in an MCDA format makes sense.

Basic Concepts of Financial Analysis

The Time Value of Money

The underlying theory of engineering economics and financial analysis is in the use of interest rate factors to find equivalent values of money. This is expressed as the time value of money, or changes in value over time, when money is either accruing interest or changing value due to inflation.

Charging interest for the use of money has its basis in capital theory, which explores whether interest should be paid, how much, and so on. Capital theory is complex, and we do not need to explore it in detail here. However, it is useful to use it as the basis for interest rate determination.

Money is human effort in storage, and if you have created capital by putting your effort in storage, you deserve a return on your savings. Say you bought a farm tractor with the money and used it to earn more than you could earn without it. The capital that went into the machine earns a return for you. If you instead put the money into a savings account, you would earn a return in the form of interest on your savings. This interest is available because the bank lends your money to someone who uses it to earn a return. The borrower pays part of that return to the bank as interest on the loan. The bank keeps some of that loan interest, and you earn some of it toward your invested capital. Everyone wins. You pay a risk premium if you put your money into secure savings rather than into a riskier business venture—the tractor.

Considering that money (or effort in storage) can be applied to produce earnings shows its value over time, or *time value*. This value is different than simply the amount of money you have now, which would be called *present value*.

Most of the basic computations of engineering economics are exercises in applying interest rate formulas to find the time value of money with different payment schedules.

Cash Flow Diagrams

The concept of cash flow is useful in financial analysis. Basically, cash flow is the movement of money in and out of an account. If no interest is charged, then tabulating cash flow is like a bank account register that shows deposits and withdrawals. Cash flow is important to show how much cash is on hand at any time and to make sure that an account is not in deficit.

Cash flow diagrams are used to show the different payment schedules. Three variables are normally displayed: present value, future value, and annual value. The simplest cash flow diagram shows present and future value. In Fig. 16-2, you can see how a present sum grows to a future value through compound interest. The diagram shows the growth of money at 2% and 8% compound interest rates. Note that the principle of compounding also applies to any growth process, such as population growth. In Fig. 16-2, you can see the dramatic effect of interest rates on compound growth. This effect is important for the growth of money and the growth of demand caused by population growth, increases in traffic, and other problems faced by infrastructure planners. In Fig. 16-3, notice how an initial amount of $1,000 can grow to more than $10,000 in less than 25 years with 10% compounding, whereas it will not even reach $2,000 if the growth is only 1%.

A "Rule of 70" (or 72) can be used to estimate doubling periods for different interest rates. If you divide 70 by the interest rate, it gives you a close approximation for the number of years to double your money. In Table 16-2,

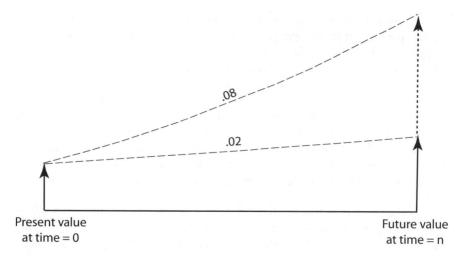

FIGURE 16-2. The growth of value from compound interest

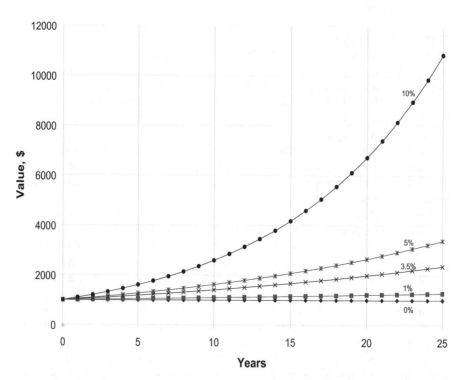

FIGURE 16-3. Compound growth as a function of the interest rate

TABLE 16-2. Application of the Rule of 70

Interest rate	Exact doubling time, years	Doubling time, using Rule of 70
0.010	69.7	70.0
0.020	35.0	35.0
0.035	20.1	20.0
0.050	14.2	14.0
0.070	10.2	10.0
0.100	7.3	7.0

you can see how well this rule works for a range of interest rates. The "Rule of 72" works slightly better for interest rates greater than about 5%.

Diagrams of compound interest show the growth over time of a sum money, but a more common situation is one where a present sum is amortized by a series of equal annual payments. The term "amortize" comes from a Latin root and means to kill off. Our English word "mortality" stems from the same root. Figure 16-4 illustrates a series of equal annual payments that will be equivalent to a present sum. An example of this kind of series is a home mortgage payment, which has equal monthly instead of yearly payments.

The Equivalence of Cash Flow and the Time Value of Money
It makes sense that a dollar today is worth more than a dollar in the future because if you invest your dollar now, its value will grow in the future. Comparing money this way entails using the concept of *equivalence*—that payments in any time pattern have an equivalent value equal to some present sum. This can be shown in the cash flow diagrams by a series of annual or future payments that are equivalent to some present value.

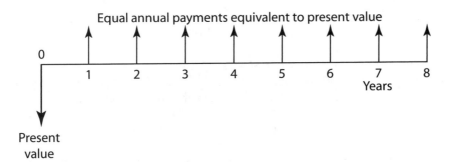

FIGURE 16-4. Equal annual payments equivalent to a present value

Equivalence is expressed by interest rate formulas, which compute payments that are equivalent to each other. Past engineering economic texts presented these formulas in detail, but today they are programmed into calculators and spreadsheets for ready use.

Usually, six formulas are presented to show all the basic situations you might want to calculate. However, you only need two of these formulas, and the other four are readily derived from the simple manipulation of these two. The exception is what is known as gradient formulas, and these will be discussed separately.

The first key equation shows a future value (F) for a present sum (P) compounded over n years at interest rate i. The factor is known as the single payment compound amount factor, or (F/P, i, n).

$$F = P(1 + i)^n$$

This was shown by Fig. 16-2, which illustrated a future value that resulted from a present investment that is compounded into the future.

The second equation computes a series of equal annual payments (A) and is very useful for a number of computations:

$$A = P\frac{i(1+i)^n}{(1+i)^n - 1}$$

This series of values was shown by Fig. 16-4, which illustrated how a present sum is equivalent to a series of equal future payments.

These two factors are convenient for a number of calculations, which have names assigned by engineering economists and which have been programmed into the Excel spreadsheet. The most common ones are shows in Table 16-3.

Applying the Formulas to Specific Problems

These formulas are simple to apply to everyday problems. Here are a few examples.

The Future or Present Value of a Single Payment
On day 1, you put $500 into an investment account that yields a nominal return of 7% per year. What is it worth in 10 years with monthly compounding?

$$F = 500 \times (1 + .07/12)^{120} = \$1,004.83$$

Note that 10 years is near the doubling time for a growth rate of 7%. You compute the present value of a single future value payment by inverting the same formula.

TABLE 16-3. Commonly used formulas for calculating cash flow and the time value of money

Name of formula	Notation	Formula	Excel function[a]
Single payment, compound amount factor	$(F/P, i\%, n)$	$\dfrac{F}{P} = (1+i)^n$	$(1+i)^\wedge n*$
Single payment, present worth factor	$(P/F, i\%, N)$	$\dfrac{P}{F} = \dfrac{1}{(1+i)^n}$	$1/(1+i)^\wedge n*$
Capital recovery factor	$(A/P, i\%, N)$	$\dfrac{A}{P} = \dfrac{i(1+i)^n}{(1+i)^n - 1}$	PMT
Uniform series, present worth factor	$(P/A, i\%, N)$	$\dfrac{P}{A} = \dfrac{(1+i)^n - 1}{i(1+i)^n}$	PV
Uniform series, compound amount factor	$(F/A, i\%, N)$	$\dfrac{F}{A} = \dfrac{(1+i)^n - 1}{i}$	FV
Sinking fund factor	$(A/F, i\%, N)$	$\dfrac{A}{F} = \dfrac{i}{(1+i)^n - 1}$	1/FV

[a]There is apparently no direct Excel function for these computations, but they are very simple. It seems logical to compute the values using these formulas, but they can also be computed using Excel functions by computing the present or future value for a series of unequal payments.

The Future or Present Value of a Uniform Series of Payments

Now you put $500 per month (at the end of each month) into the investment account that yields a nominal return of 7% per year. What is it worth in 10 years?

$$F = 500 \times (F/A, i\%, n) = \$86,542.40$$

If the problem was to find what this future sum required in monthly payments, you would simply invert the F/A formula to find A/F.

If you choose to compute the present value of this investment, the equation is:

$$P = 500 \times (P/A, i, n) = \$43,063.18$$

This sum, converted to a future value, is:

$$F = P \times (F/P, i, n) = 43,063.18 \times (1 + i)^n = 43,063.18 \times (1 + .07/12)^{120} = \$86,542.40$$

Note that the sinking fund, or the amount you need to "sink" into an account to pay for something in the future, is simply (A/F, i, n).

There are many other examples of how to use these formulas, and they can be reviewed in texts such as that by Grant and others (1997).

Gradient Series

As mentioned above, the basic list of formulas is for single payments or equal annual series. In the event a series of payments increases over time, either arithmetically or geometrically, it is known as a gradient series. An example of an arithmetic gradient is shown in Fig. 16-5. Notice that the gradient series is composed of a series with equal payments and another one that begins at time 1 with zero payment and increases linearly. The equation for the increasing portion is:

$$A = \frac{G}{i} - \frac{nG}{i}\frac{i}{(1+i)^n - 1}$$

This equation computes a uniform payment that is equivalent to the part of the gradient shown in Fig. 16-5. You simply have to add it to the other uniform component to get the total uniform equivalent payment.

Evaluating Alternative Investments

When you are evaluating alternative investments, you apply attributes such as those advocated by Grant and others (1997), which include clear alter-

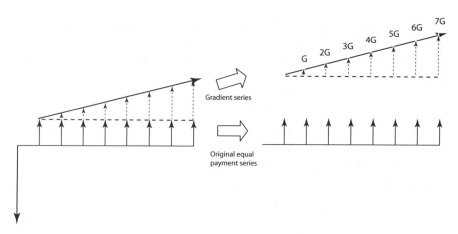

FIGURE 16-5. A gradient series composed of equal series and a pure gradient

natives, the merits of all appropriate alternatives, the expected future consequences, the viewpoint for weighing merits, differences among alternatives, nonmonetary consequences (intangibles), and side effects. These apply across the board for decisionmaking, in both the private and public sectors.

An example of comparing alternative public sector investments would be to rate one package of transportation capital improvements versus another, for instance, a wider road against improved traffic controls. Another example would be one water resources project versus another, for instance, a larger dam versus a smaller one plus channel improvements. In the private sector, investments might be for business expansion purposes. The evaluation of public sector investments requires consideration of more side effects and intangibles than is normally the case in the private sector.

To formulate alternatives, you would package mutually exclusive projects (Plan A, Plan B, etc.) or compile combinations of alternatives. The "do nothing" alternative should also be included to create a basis for comparison.

Although engineering economics offers several ways to display numbers to compare alternatives, there are only two basic ways. In one, you compute the economic parameters, such as present value, future value, annual value, and the rate of return of the net cash flow. In the other, you perform "benefit-cost analysis."

In these kinds of evaluations, you must specify the interest rate as your "minimum attractive rate of return" (MARR). This is sometimes hard to do, but if you use the cost of money—based on an indicator such as the Federal Reserve's charge to member banks, the rate for long-term Treasury bonds, or the prime interest rate—you will be consistent. Rates like these are logical, because if you did not invest in your infrastructure project, you could get at least those rates for your funds if you invested in the alternative investments.

Net Cash Flow

Net cash flow analysis is a way to display the time variation of costs and revenues. Say you build a toll road for $20 million and it costs $2 million per year to operate and maintain. Your analysis period is 25 years. If the MARR is 7%, then the net discounted costs are:

Present value of all costs: $43,307,166
Future value of all costs: $235,046,728
Annual value of all costs: $3,716,210

The interpretation of these numbers is that the actual annual cost of the road is $3.7 million for capital and operations. If you capitalize the operations and maintenance and add it to original cost, the present value is $43 million, and if you extend this to its commensurate value 25 years into the future, it comes to $235 million. These numbers might be used to help determine the size of a bond issue to pay for the road's construction and to subsidize operations and maintenance until revenues climb to the point where they will cover both capital and interest.

Benefit-Cost Analysis

Benefit-cost analysis is appropriate for problems of the public sector because it is a flexible procedure and you can consider different categories of benefits and costs. It goes well with multiobjective analysis when you are considering environmental and social costs. BCA has many specific requirements, which are explained in textbooks such as that by Gramlich (1990).

As mentioned above, it is usually reported that BCA had its formal beginning with the Flood Control Act of 1936, giving the water resources field credit for its initiation. It seems likely that the drafters of that bill drew on thinking about benefits and costs that preceded the bill itself, however. Whatever its origin, the BCA concept is to array all benefits and all costs and to compare them using consistent criteria.

For example, say you have a project that will return benefits of $1 million per year over its lifetime of 20 years, and say you can achieve these benefits with a present investment of $8 million and annual operating costs of $200,000. In this case, the benefits are the annual $1 million returns. The costs are the initial capital cost of $8 million and the annual operating costs of $200,000. If the interest rate is 7%, then the computations are as shown in Table 16-4. These results were quickly generated using an Excel spreadsheet.

This analysis also reveals how you can compute the rate of return. You would insert different interest rates until the benefit-cost ratio = 0. Using a trial-and-error method, this turns out to be 7.75%. Using the Excel function for the internal rate of return, the answer is 7.7547%.

Inflation

Financial studies require consideration of the effects of inflation, which can be very significant. Though inflation has many impacts on the economy,

TABLE 16-4. Sample benefit-cost analysis

	Data	Present value	Annual value
Investment	$8,000,000	$8,000,000	$755,143
Operating costs	$200,000	$2,118,803	$200,000
Returns	$1,000,000	$10,594,014	$1,000,000
Interest rate	0.07		
Years	20		
Net benefits		475,211	44,857
Benefit-cost ratio		1.05	1.05

its main effect in the infrastructure and environmental areas is a loss of the purchasing power of money as measured by price indices.

Say you have a fixed sum of money available at $1,000,000 and must build a project that costs $950,000 today. Inflation is at 4% per year. If you delay your project, inflation erodes the purchasing power of your money, as shown in Table 16-5. So, if you delay until after year 1, you will not have enough money to do the project—unless, of course, you are able to invest your $1,000,000 in an account that compounds interest so you can keep up with the inflation loss. If you can earn more money than inflation, you can even build up your funds and build a bigger project. That is why the advice you get is to borrow and build your project bigger now if you expect inflation to be high.

Loan Amortization

Using the capital recovery factor (the Excel *PMT* function) offers a lot of information about the cost of money. To see this, set up a spreadsheet that includes the information in Table 16-6. This one is shown for only 10 years to save space, but it could be set up for a 30-year house payment with 360 monthly payments and a monthly interest rate equal to 1/12 of the annual rate. The table is self-explanatory. The equal annual payments are made at the end of each year. The interest during each year is the interest rate times the balance remaining. The principal that goes toward repaying the loan is the difference between the payment and the interest. The balance due goes exactly to zero at the end of year 10. The sum of principal payments is the original loan amount.

Now, having this spreadsheet enables us to make useful computations, such as to analyze the savings in interest if additional payments are made on the loan principal. Using a longer-term loan as an example, Fig. 16-6 illustrates the principal and interest payments and the balance if the original payment schedule is followed and if accelerated payments are made. Notice on the graph that the interest payments fall much more quickly if the

TABLE 16-5. The future value of $1 million today, with inflation

Year	Value
0	$1,000,000
1	$961,538
2	$924,556
3	$888,996
4	$854,804
5	$821,927

TABLE 16-6. Sample information to calculate loan amortization

Loan	1,000,000
Interest	0.07
Periods	10

	Payment	Interest	Principal	Balance (last day of the month)
0				1,000,000
1	142,378	70,000	72,378	927,622
2	142,378	64,934	77,444	850,179
3	142,378	59,512	82,865	767,314
4	142,378	53,712	88,666	678,648
5	142,378	47,505	94,872	583,776
6	142,378	40,864	101,513	482,263
7	142,378	33,758	108,619	373,644
8	142,378	26,155	116,222	257,421
9	142,378	18,019	124,358	133,063
10	142,378	9,314	133,063	0
Totals	1,423,780	423,773	1,000,000	

payment is increased. By the same token, the principal payments rise much more quickly. This is the basis for your incentive to repay loans quickly to save on the total interest paid.

Cash Flow Analysis

One of the most useful financial tools is cash flow analysis. The term "cash flow" has specific meanings within accounting to measure the cash received and spent within a period of time. The way I am using the term here is to refer to the cash inflows and outflows by period for some enterprise. Perhaps the best way to illustrate this is with an example. The following cash flow scenario is for a hypothetical problem where you analyze the cash flow and projected financial statements for a future water system expansion for a small town. The data given is outlined here:

The water system would be expanded to handle a population influx of 5% per year for the next 10 years. Half the expanded system would be built now, and half in 5 years. Funding for construction would be from loans

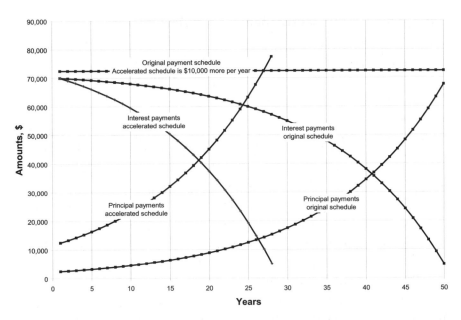

FIGURE 16-6. The loan amortization process, showing accelerated payments

from an infrastructure bank. Annual loan payments begin at the end of year 1 and continue for 10 years for all loans. Land use is mixed residential and commercial, but for simplicity we consider only residential. The data are presented in Table 16-7, and a simplified cash flow analysis of the data is presented in Table 16-8.

Even this simple problem becomes complex because the numerical data quickly increase. The intent is to show the inflows and outflows of cash for each year. In this case, the revenue consists of receipts for water sales, tap fees, and sales and property taxes. Total costs are shown, and net revenue is positive. Because net revenue is so high, it suggests that rates or charges should be cut or that the loans should be repaid more quickly. One problem not apparent from Table 16-8 is that the rate of system renewal is too low, and deferred maintenance is building up too quickly. Therefore, the surplus funds might need to be invested in system renewal.

Transportation Analysis Examples

Transportation problems offer a good case study of use of engineering economics or financial analysis to compare alternative investments. Two

TABLE 16-7. Sample data needed for a cash flow analysis

Current population	100,000 (33,333 households)
Rate of population growth	5% per year for 10 years; 0% after that
Per capita water usage (average)	150 gallons per capita per day
Land use	Mixed residential and commercial
Planning horizon for capital improvements	10 years to meet demands; 30 years for system life
Plant investment fee	$5,000 per house connection
Current water fees	$2.50/1,000 gallons
Property tax levy for water system improvements	0.7 mills
Assessed valuation of current residential property	$980 million (figured as market value × 0.2)
Sales tax dedicated to water system	0.8%
Current anticipated taxable sales	$500,000,000
Interest rate on loan (due in 10 years)	8%
Projected inflation rate	0%
Capital cost of system/replacement value of old	$3,000 for each new person
Current value of existing system (age 15 years)	Replacement value less 15 years at 3.33%
Capital improvement goal	Current system value at least 50% replacement value
Depreciation of assets	3.33% per year
Operations and maintenance cost	$50 per capita per year

examples are presented here. One is to compare a financial investment with the economic benefits of a transportation system, and the other compares different ways to obtain funds to improve a road system.

Example: The Economic Benefits of a Transportation System

A rural highway project costing $20 million will flatten slopes, reduce curves, and shorten a road from 12 miles to 11 miles. These improvements will reduce average travel times by 5 minutes, reduce traffic accidents by an average of 5 per year, and reduce annual maintenance costs by $10,000. Average

TABLE 16-8. Cash flow analysis of the data presented in table 16-7

Year	Population	Revenue	Loan payment	Operations and maintenance	Depreciation	Total cost	Net revenue
0	100,000						
1	100,000	23,106,833	4,313,619	5,000,000	4,995,000	14,308,619	8,798,215
2	105,000	24,262,175	4,313,619	5,250,000	5,792,526	15,356,145	8,906,030
3	110,250	25,475,284	4,313,619	5,512,500	5,599,635	15,425,753	10,049,530
4	115,763	26,749,048	4,313,619	5,788,125	5,413,167	15,514,911	11,234,137
5	121,551	28,086,500	4,313,619	6,077,531	5,232,909	15,624,058	12,462,442
6	127,628	29,490,825	8,627,237	6,381,408	5,058,653	20,067,298	9,423,528
7	134,010	30,965,367	8,627,237	6,700,478	5,854,059	21,181,774	9,783,592
8	140,710	32,513,635	8,627,237	7,035,502	5,659,119	21,321,858	11,191,777
9	147,746	34,139,317	8,627,237	7,387,277	5,470,670	21,485,185	12,654,132
10	155,133	35,846,283	8,627,237	7,756,641	5,288,497	21,672,375	14,173,907

daily traffic (ADT) is currently 10,000, with a load factor of 1.5 persons per vehicle. Traffic will increase linearly by a total of 30% over the 20-year life of the project. If vehicle operating costs are $0.15 per mile, travelers value their time at $5 an hour, and the average cost of a traffic accident is $20,000, what is the benefit-cost ratio of the proposed investment? Assume the interest rate is 8%.

Solution

The problem can be solved several ways. The benefits are the sum of annual savings for traveler time, operating costs, maintenance costs, and accident reduction. The costs are the annual cost of the initial investment. If the operations and maintenance costs were included on the cost side, the results should be the same.

Given that average daily traffic increases by 1.5% per year, you must consider this increase over the life of the project to obtain the equivalent annual benefit for traveler and operating cost savings. This could be computed using a gradient formula or by summing the annual values on a spreadsheet and discounting them to the present and annual values.

Table 16-9 provides the values from the solution.

Notice that the compounded growth in ADT requires you to increase the base year values for benefits for traveler savings and vehicle operating cost savings. This means that you get future benefits because of this growth. The value of these future benefits is sensitive to two rates, one for annual growth in traffic and the other for the interest rate, which is used to discount the value of the future benefits. A third rate, the inflation rate, might also enter the equation, but we ignored inflation in this analysis. If you think about the uncertainties of projecting rates like these 20 years into the future, you see how economic analyses can be wrought with uncertainty about the future.

Example: Alternative Ways to Obtain Funds to Improve a Road System

The first example was about economic costs and benefits. This one deals with the cost of money. Say you can obtain your $20 million for investment by borrowing, creating a sinking fund, or selling your assets to the private sector. In the case of the above example, only borrowing makes sense, because you need to build the project now and selling the road is not practical. A sinking fund would require you to wait too long to build the project. So we consider two ways to borrow the money: a bank loan or selling bonds.

The bank loan at 7.5% can be repaid in equal annual payments over 20 years. In addition, the bank will charge a 1.5% setup fee. The bonds can also be spread over 20 years and, because they will be tax-exempt municipal bonds, will carry a coupon rate of 6%. The underwriter will charge 2% to handle the transactions.

TABLE 16-9. Benefit-cost analysis of a sample transportation system

Project cost	$20,000,000
Term in years	20
Travel savings per traveler, minutes	5
Accident reductions per year	5
Maintenance savings per year	50,000
Average daily traffic, base year	10,000
Load factor	1.5
Increase in average daily traffic per year	0.015
Operating cost per mile	$0.15
Value of traveler time	$5.00
Cost per accident	$20,000
Interest rate	0.08
Annual costs	
Investment	$2,037,044
Annual benefits	
Traveler savings, base year	$2,281,250
Accident prevention	$100,000
Vehicle operating savings, base year	$547,500
Maintenance savings, per year	$50,000
Compounded traveler and operating savings	$3,403,817
Total annual value, benefits/savings[a]	$3,553,817
Benefit-cost ratio	1.74

[a]Computed as the sum of compounded traveler and operating savings, accident savings, and maintenance savings.

Solution

This calculation was done on a spreadsheet. It showed that if you wrap the 1.5% fee into the bank loan balance and amortize the resulting loan of $20,300,000 over 20 years at 7.5%, the annual payment is $1,991,271 and the total over 20 years is $39,825,430. The bond payment is 6% of the original proceeds, and if you pay the 2% fee in the first year and retire the full $20 million in bonds at the end of the twentieth year, your total bond payments are $43,200,000.

Thus, the bonds actually cost more over the 20-year term, even though they carry a lower interest rate. This happens because as you pay off the bank loan, your interest payments drop as you pay off the principal. With the bond, you do not pay off the principal until the end, so you have more capital at your disposal.

Analyzing Risk and the Cost of Capital

Many other techniques are available for financial analysis. In Chapter 10, the use of analysis to probe financial statements and compute ratios was an example. Two other examples are the use of a capital asset pricing model to compute the cost of capital and the use of the beta factor to compute the risk of stock market investments.

The Cost of Capital

The "cost of capital" determines if a capital budgeting decision makes sense. Whereas capital budgeting for public infrastructure is mostly done by considering political priorities, private businesses examine their priorities on the basis of business decisions.

The concept of cost of capital is based on debt (obligations to lenders) and equity (obligations to owners). A variable labeled the "weighted average cost of capital" has been developed to measure the cost of capital, considering debt and equity (Block and Hirt 1997). It enables you to determine if management is investing its capital well by comparing it with another variable, return on investment capital (ROIC), which is defined as:

$$ROIC = (\text{net income} - \text{dividends})/\text{total capital}$$

Risk: The Beta Factor

To measure risk in investments, the Capital Asset Pricing Model (CAPM) can be used (McCracken 2005). It compares the risk/return trade-offs of individual investments to average market returns. The CAPM recognizes that investors need to be compensated for the time value of their money and the risk they take. The time value of money can be represented in a formula by the "risk-free" rate (rf). The formula can also include risk to estimate the compensation the investor deserves for taking on risk. Risk is calculated by the "beta" measure that compares the returns on the asset with those of the market over a period of time and with the market premium (Investopedia 2007).

The equation that correlates these is:

$$Ks = Krf + B(Km - Krf)$$

where:

Ks = rate of return to justify having a stock with this beta
Krf = rate of return on a risk-free investment like U.S. Treasury bonds
B = Beta, computed as the rate of return of the stock versus the rate for the market
Km = expected return on the overall stock market

For the overall market, $Ks = Km = Krf + B(Km - Krf)$, so beta = 1.0.

Financial Engineering

The term "financial engineering" has been coined to refer to the use of financial instruments to structure portfolios to improve their performance (Galitz 1995). Financial engineering is a tool of finance more than engineering. The term "engineering" conveys the concept that mathematical analysis and deliberate design are used to construct strategies to improve financial portfolios. The goal of this chapter is to present a simple and concise summary of financial engineering as it might affect civil engineering, construction, and public works managers.

Since 1992, financial engineering has had its own professional association, the International Association of Financial Engineering (IAFE, www.iafe.org). Its mission is to foster quantitative finance among academics and practitioners from banks, broker dealers, hedge funds, pension funds, asset managers, technology firms, regulators, accounting, consulting and law firms. Mathematics departments of some universities have created programs in financial mathematics and financial engineering.

IAFE has six committees, whose missions outline its view of the field. The Credit Risk Committee focuses on risk at the single obligation and portfolio levels and in institutional settings, as well as on risk's policy implications. This might involve computational analysis of asset pricing, portfolio management, and risk control. IAFE's Education Committee focuses on a knowledge core of financial mathematics and financial engineering by working with human resources staff, business managers, and educators. The topics it addresses include financial theory, mathematics and statistics, computing, and business knowledge.

The IAFE Investor Risk Committee includes fund managers, institutional investors, regulators, brokers, consultants, custodians, technology vendors, and others who work in the fund arena. It has produced a white paper,

"Valuation Concepts for Investment Companies and Financial Institutions and Their Stakeholders." It studies disclosure for hedge funds.

Next is the IAFE Liquidity Risk Committee. Liquidity risk is a threat to the function and stability of financial markets and institutions. Financial intermediaries (banks, broker-dealers, insurers, pensions, and asset managers, hedge funds) are becoming more market oriented, and the effect of liquidity risk upon their cost structure and access to credit is evident. Liquidity risk is linked to such phenomena as asset bubbles and market crashes and undermines price efficiency and market behavior.

The IAFE Operational Risk Committee addresses best practices, quantification, corporate governance, technology applications, the effectiveness of existing controls, and operational risk's interrelationships with other forms of risk. This committee addresses breakdowns or failures relating to people, internal processes, technology, and the consequences of external events— encompassing the risk inherent in business activities across the firm. The committee believes that operational risk is itself a broad discipline, and it includes business reputation, strategy, and corporate governance in its definition. It provides a forum for major business and technology paradigm shifts in the financial engineering industry.

Finally, the IAFE Committee on Financial Modeling and Technology is working on valuation and modeling tests for credit derivatives and other emerging products, data and interface standards for Web-based risk management and reporting, financial models and derivative product definition standards, best practices for information technology architecture to support risk measurement, and emerging techniques for electronic processing of financial instruments.

Applications to Infrastructure and the Environment

Engineering economics and financial analysis are among the most useful tools for those who manage infrastructure and environmental systems. They start with the big picture—using methods such as multicriteria decision analysis, benefit-cost analysis, and environmental and social analysis— and they extend to quantitative approaches using principles of engineering economics and financial analysis along with formulas for the time value of money.

As the following chapter shows, today's approaches to managing infrastructure and the environment focus on getting more return for the dollar, and this requires a very sharp pencil to evaluate alternative strategies. This sharp pencil is furnished by the tools that have been presented in this chapter, which show how to use cash flow analysis to evaluate scenarios such as finding the best ways to raise capital and lower project costs while increasing rates of return. This chapter has also explained perplexing phenomena such as inflation, how to analyze risk and the cost of capital, and the new field of financial engineering.

References

Block, Stanley B., and Hirt, G. (1997). *Foundations of financial management.* 8th ed. Irwin, Homewood, IL.

Galitz, L. (1995). *Financial engineering: Tools and techniques to manage financial risk.* Financial Times / Prentice Hall, London.

Gramlich, E. (1990). *A guide to benefit-cost analysis.* 2nd ed. Prentice Hall, Englewood Cliffs, NJ.

Grant, E., Ireson, W., and Leavenworth, R. (1997). *Principles of engineering economy.* 8th ed. John Wiley & Sons, New York.

Hager, W. H. (2004). "Jules Dupuit: Eminent hydraulic engineer." *Journal of Hydraulic Engineering,* 130(9), 843–48.

Hill, M. (1968). "Goals achievement matrix for evaluating alternative plans." *Journal of American Institute of Planners,* July.

Investopedia. (2007). CAPM. www.investopedia.com. Accessed September 2, 2007.

McCracken, M. E. (2005). CAPM. http://teachmefinance.com/capm.html. Accessed July 16, 2005.

Newnan, D. G., Lavelle, J. P., and Eschenbach, T. G. (2004). *Engineering economic analysis.* 9th ed. Engineering Press, San Jose, CA.

Stanford University. (2006). Department of Management Science and Engineering. http://www.stanford.edu/dept/MSandE/about/history.html. Accessed May 21, 2006.

U.S. Water Resources Council. (1983). *Economic and environmental principles and guidelines for water and related land resources implementation studies.* U.S. Water Resources Council, Washington, DC.

Wellington, A. (1887). *The economic theory of railway location.* 2nd ed. John Wiley & Sons, New York.

17

New Tools for Managing Infrastructure and the Environment

Economics and Finance for Infrastructure and the Environment

In today's multifaceted world, the complexity of infrastructure and pressure on the environment require better tools to manage and sustain them. Gone are the days when the engineering problem was formulated as "Determine the need and build a system to meet it." Today, many issues enter the equation, especially using demand management and low-impact systems. The "triple bottom line" metaphor captures the sustainability goal: to meet economic, social, and environmental needs in a balanced way.

Economic and financial tools offer bright hope for new and innovative solutions. This book has explained how economic tools apply to land use, transportation, construction, and utility systems, while protecting environmental resources. The second part of the book explained how tools for accounting, public finance, capital markets, asset management, and decision analysis lead to practical financial solutions for infrastructure and the environment.

This chapter introduces a set of concepts that converge to create a stream of emerging ideas about how economics and finance can help sustain infrastructure systems and environmental quality, thus supporting a vibrant civilization with its natural systems intact. These concepts provide solutions to the challenge of how to balance command and control, the free market, and idealistic visions about the future. The tools that are coming on line will enable us to recognize limits and allocate resources to achieve "triple bottom line" objectives. New technologies will help a lot in the future, but the most exciting possibilities will be in new management solutions.

The Imperative: Improving Life and Overcoming Limits

Everywhere you look, society faces limits. There are an infrastructure crisis, an energy shortage, pollution of the environment, congested roads, and global warming. Some pundits say: "We have reached the limit, and the only way to a sustainable future is to cut back." Others take the opposite view: "There is nowhere to go but up, because technology will solve all our problems and we can have a better life without ruining the environment."

Although neither idea is exactly on target, almost everyone is interested in both growth and sustaining the environment. People want a better life, but they do not want to ruin natural systems. The question is how we do this. This chapter is organized around answers to this question, with emphasis on:

- finding the balance between government and private sector management of infrastructure;
- privatizing services and facilities that are appropriate for the private sector;
- managing demand for public goods through better pricing systems;
- reducing the cost of government while improving public services;
- deregulating costly industries where monopoly services rule; and
- meeting social needs while pricing public services and environmental systems.

In Chapter 2, I addressed the balance between government and the market economy. We considered "market failure," where private sector activity does not meet the public's needs efficiently, and "government failure," for example in a case with high taxes and fees and inefficient services. I explained how, in the mixed U.S. system, three methods of delivering public services are in use. The traditional public model is where a government department provides services, for example, in a city light and power department. Privatization is the model where a service is offered by the private sector, either as a regulated monopoly, such as a public company distributing electric power, or through competition, as when different companies compete to collect solid wastes. The "in-between" model has different names. One that

I explained is "managed competition," which allows government services, privatization, or a mixture.

Traditional arguments for government monopolies are that a service is "not appropriate" for the private sector, that it is too risky for a private company to offer the service, that a government department can care for the environment better while exploiting it for infrastructure services, and the like. These and related arguments were explained in Chapter 7, which covered utility economics. It showed how, in general, three types of goods might be considered for government or private provision:

■ public utility–type goods that are essential public services that can be measured and rationed by charging schemes;
■ Private goods with important public purposes or those that provide benefits to society as a whole but can be offered by private firms; and
■ Services where public purposes dominate but one person's use of the service does not diminish its availability to others.

In examining the arguments for and against government involvement, we considered five variables: whether the service is essential, whether it requires regulation, whether use by one person diminishes availability to others, whether it must be a monopoly, and whether it is really a commodity service. The public is getting sharper about answering these questions for infrastructure services. Each of the services discussed in the book (the built environment, energy, water, transportation, communication, and waste management) has elements that can be unbundled and deregulated, thus opened for the private sector without government monopolies. The question is "How to do this?" In most cases, creative legislation is the answer. This involves risk, because we simply do not know what will happen until we try.

So balancing government and private provision of infrastructure and public services requires the will to do so, the knowledge to unbundle and deregulate services, valid pricing systems, and a residual method to deliver nonmarket vital services, including those with a required social component. The sequence to analyze this begins with unbundling services. This leads us to answers to questions such as "How much does the public need this service?" "Is it essential that government provide it?" "What parts of it can be privatized or outsourced?" and "How should we use pricing of this service to allocate the public's resources?"

Unbundling Services

To facilitate competition and to make pricing and privatization work, services must be unbundled. This requires doing a careful analysis of how they are bundled and of the costs of different parts of a bundled service. This means that financial accounting for parts of organizations must be precise.

Identifying parts of a service is similar to identifying business processes in any operation. The question is "What processes are necessary to produce the total, integrated service?" When you list the separate processes, you are, in effect, unbundling them. For example, electric power involves generation, transmission, and distribution. If the service is vertically integrated, one organization charges for all. If these are unbundled, the charges can be separated. This is actually a key feature of the deregulation of electric power. Generation can be by anyone. Transmission can be open for competition. Distribution is a natural monopoly and cannot be unbundled. Of course, there are many other business processes in electric power services, and these are just the main system processes.

Transportation is a natural candidate for unbundling because the different modes are already separate, but their joint use of public roads and spaces offers opportunities to separate components of services and to identify user costs. To explain further, the concept of "transportation" is a high-level system, and identifying its sectors constitutes an unbundling of sorts. When you examine a lower-level system, such as a city street, and you look at what it takes to operate and maintain it, you see many other opportunities for unbundling. For example, traffic control, street cleaning, maintenance and renewal, drainage, and other elements are required to keep it in operation.

Of course, many other aspects of government services can be unbundled. Sometimes we do not even notice them as, for example, in letting private sector contracts for the maintenance of public buildings. As another example, an urban drainage system can have maintenance unbundled from other services. The unbundling exercise does not only look at systems and business processes; it also looks at how an organization is structured to perform its overall mission.

Deregulation

The term "regulate" means to control behavior in accordance with a rule or law, and it is aimed at protecting the public interest where private markets do not. Many examples can be cited: the sale of alcohol, highway speed limits, the practice of medicine, and so on. Regulators are an important part of the infrastructure and environmental industries, and they enforce rules about health and safety, environmental quality, service quality, and finance.

Regulations are developed to implement laws, and the Administrative Procedures Act was passed to provide agencies with guidance in rulemaking. For example, the Safe Drinking Water Act was passed by the U.S. Congress and signed by the president. Under the authority of the Safe Drinking Water

Act, the Lead and Copper Rule is a regulation issued by the Environmental Protection Agency.

Regulatory programs should follow the principle of "not to have the fox guarding the chicken coop." This recognizes that persons should not be expected to regulate themselves. Conversely, the same agencies that write the rules enforce them, so regulators need oversight as well. These are examples of why the "separation of powers" is required in government.

The regulatory arena is where conflicts over business versus environment are worked out. In this sense, regulation is a "coordinating mechanism" for industry. Interest groups push their agendas through regulations and laws.

A regulatory program must have an enforcement mechanism to be taken seriously. Like officials at a sports event, enforcement staff should know the rules well. Officials must have reliable information for making decisions. Enforcement officials should try to obtain compliance before levying penalties. The system of enforcement must be efficient. Enforcement must be fair, and appeals panels must be available to provide due process, as well as to back up the regulatory goals.

The total picture of infrastructure and environmental regulation is a mixture of federal, state, and local laws and regulations that govern service providers. Because much of infrastructure service is by local government, regulation comes from federal laws, implemented by state agencies. Other regulation is informal, through the political process. For example, rate setting by local governments normally requires no approvals, whereas rate setting by private utilities is regulated by public service commissions.

Calls for "regulatory relief" and "regulatory reform" are common because people and businesses don't like being regulated. However, regulation is a price to pay for civilized society. The challenge is to regulate enough but not too much. Regulation seeks to apply laws to control behavior in the public interest, but defining the public interest is an elusive goal.

The reasons that a service is regulated include:

- health and safety (e.g., road design for safety);
- environmental quality (e.g., maintain clean air);
- resource allocation (e.g., recognize legal water rights);
- finance (e.g., control charges by airlines); and
- service quality (e.g., require solid waste collection services).

Once a service is unbundled, the need for regulation of each piece can be studied. If desirable, a piece can be deregulated. Obviously, this is complex and involves many aspects—take, for example, the deregulation of electric power. First, it is unbundled into generation, transmission, and distribution, and the electric power companies are required to divest so as to facilitate competition in these parts of the industry. If generation, for example, is deregulated, then no longer is a company required to have a certain level of capacity but instead the market is opened for other players to enter and generate power, with a level playing field to sell it. Regulation is now of a different character and meant to facilitate competition and improve service.

Privatization

The question of whether all or part of a service can be privatized is different than the one about regulation. Given that infrastructure and environmental protection have strong public purposes, we need to know what is appropriate for private sector provision. To do that, we can examine the concepts of, arguments for, and track record of privatization.

The privatization of infrastructure and public services became hot about 1980, when the end of the Cold War and opening of socialist countries to competition fueled interest. Over the years, many arguments for privatization have been made, and quite a bit of it has been tried. Many times it has turned out to be less favorable than the advocates thought. But on the whole, it has been a good thing. Actually, the "third sector" of nongovernmental organizations can meet many needs. Their efforts go beyond for-profit business or government efforts and seek to use good will and private efforts for work toward the common good.

There are, of course, many examples of privatization, and the list includes familiar arrangements, such as private railroads, transit systems and intercity bus lines, private water companies, wastewater service privatization, solid waste collection, private ownership or operation of public buildings, and private sector electric and gas companies.

The arguments for privatization center on using private sector efficiency to overcome public sector problems. In many ways, it is the same as the argument as to whether capitalism or socialism is better. A quote by a British official explains the idea: "Competition is an extraordinarily efficient mechanism. It ensures that goods and services preferred by the consumer are delivered at the lowest economic cost. It responds constantly to changes in consumer preferences. It does not require politicians or civil servants to make it work" (Walker 1985).

Figure 17-1 shows a matrix with a public-private axis to move between capitalism and socialism and a vertical axis to move from an emphasis on efficiency to an emphasis on equity and the public interest. Two quadrants of the matrix are easy to identify: for-profit business for the sale of any private good or service, and nonprofit nongovernmental organizations to offer social programs. The other two are not as easy to restrict to one entity or another.

Utility and some public services can be offered by either the private sector or by government and, though they focus on efficiency, they also deal

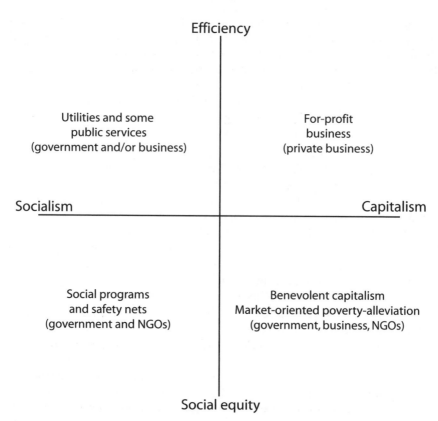

FIGURE 17-1. Arenas for activity by government, the private sector, and nongovernmental organizations (NGOs)

with some equity issues. Chapter 7 explained the issues involved here. The lower-right quadrant is for capitalism with social purposes and, although it lacks many successful examples, some are reported in developing countries, where small loans by a nongovernmental organization, for example, can spark individual initiative and small business successes.

What is appropriate for government is being studied at many places, including the Department of City and Regional Planning at Cornell University (2007). These researchers explain how privatization is occurring worldwide at all levels of government with the goal to reduce costs by turning government services over to the private sector. The appeal of privatization is the belief that market competition is the most efficient way to provide services and allows for greater citizen choice. They acknowledge concerns about service quality, social equity, and employment conditions.

The Cornell researchers also explain movements to improve the efficiency of public administration as they are sparked by privatization and business-model prescriptions for government. The spectrum goes from those who see government as a business that provides services to citizens at the lowest possible cost to those who believe that government should focus on public values such as equity, accountability, and citizen voice. Table 17-1 summarizes the arguments for and against privatization.

History and Track Record of Privatization

So, given that privatization heated up in the 1980s, what is its track record after more than two decades? The British example of selling off state-owned enterprises during Margaret Thatcher's government attracted much attention. They sold many companies, including British Aerospace, Cable and

TABLE 17-1. Sample arguments for and against the privatization of government services

Type of argument	For privatization	Against privatization
Ideology	Water is an economic good and ought to be managed by the private sector; smaller government is better	Water is a public good and ought to be managed by government; government is needed to resist excesses of the private sector
Efficiency	The private sector is more competent than the public sector; cost savings in construction, procurement and management, in hiring and training; tax benefits	The public sector is efficient; cost savings in privatization are fictional; tax benefits are a shell game; negative aspects of long-term contracts; the potential for rate increases
Social impacts	Government is not reliable and people do not always trust it	Loss of political control; loss of jobs; loss of public benefits of water
Risk reduction	Private firm will guarantee performance	Effective management by government minimizes risk
Capital generation	Access to capital in private markets; government debt limits not imposed	The public sector should generate capital for infrastructure

Wireless, Amersham International, Britoil, and Associated British Ports. Future targets for privatization were the Electricity Council, British Telecom, British Gas, British Steel, BL, British Rail, British Airways, Rolls-Royce, British Shipbuilders, National Bus, Royal Ordnance Factories, and the British Airports Authority. They also privatized their water industry in the 1980s (Brown 1983). On the whole, this seems to have been successful, judging from the increased competitiveness of British business now.

The U.S. experience has been different from that of the British because America is much larger and more varied by state and region. One focus has been on deregulation, with the highest-profile move being to deregulate the national telephone monopoly. The nation has, in addition, many private sector energy and water companies. A number of local experiments with the privatization of water, wastewater, and other public services have been carried out. The scorecard of success is mixed on these.

Developing countries have networks of state-owned enterprises to provide services, but many lack a good track record. Many have turned to privatization as an alternative to state companies. Again, the record is mixed. Some successes have been recorded, but public protests against privatization in some places have turned violent. People are interested in their social safety nets and often fear privatization.

Thus, it is hard to generalize, and the conclusion is that privatization has a mixed record. In the United States, it has good potential for many applications but not everything. Transportation infrastructure is an area for further exploration. As explained by Butler (2007), the United States is dependent on interlocking infrastructure networks and nodes, but their deterioration puts the nation's economic competitiveness in jeopardy. Gasoline tax funds to finance the Interstate Highway System have not been indexed to inflation and are diluted by earmarks, which reached $23 billion in 2005. Without reform, financial shortfalls are likely, causing system deterioration and more congestion. Elected officials lack incentives to raise tolls or taxes, inefficiency results from a lack of competition and market forces, and politicians respond to politics, not the national interest. Privatization might provide capital, increase construction and operational efficiency, transfer risk to private investors, and create new sources of tax revenue. If private entities could build profitable roadways, federal funding could be concentrated on other vital projects.

Pricing Public Goods to Manage Demand

Whether infrastructure and public services are privatized or not, we face the question of how to price them. Chapters 7 and 12 explained the theories of utility financing, but pricing can be used to manage demand and achieve beneficial public purposes in ways that go beyond charging for services and recovering costs.

These ideas are already encapsulated in the principles of rate setting delineated in Chapter 12, but they go beyond the basic concept of "the user pays" to emphasize the need to ration the use of a facility or service. The principles that most apply are the user pays, to ration the use of a service, and to apply peak-load pricing.

A few examples can illustrate the possibilities. Take the old concept of the toll road, for example. If tolls are constant and set at the cost of service, then the toll takes on the feature of a cost-recovery device. If, however, you extend the concept to charging tolls for special services such as a high-occupancy lane, then you have the possibility of pricing a service. These high-occupancy lanes are available to offer drivers the option of selecting a higher level of service for a price that varies with the degree of congestion.

A related device is to charge drivers by the time of day that they take a road. If they drive during peak hours and contribute to congestion, the cost is higher than if they drive earlier or later. The incentive is thus to drive earlier or later, thus flattening the peak of congestion. This is an example of congestion pricing, and the same concept can be used to balance electric power loads by time-of-day pricing. Electric power requires a meter that records not only how much power is used but when.

In the fields of water and energy, conservation rates can be used to price the commodity. In water pricing, for example, the traditional cost-of-service rate structure trended downward, but a conservation rate does just the opposite (Fig. 17-2). With the cost-of-service rate, there is no incentive to conserve. The incentive is just the opposite, to use more. But with the conservation rate, you get a basic quantity for a low rate, and you pay more as your use increases.

Other new and innovative ideas are coming on line as well. For example, the City of Boulder (2007) has gone to the "water budget" rate structure. In it, you are entitled to a basic amount of water for indoor use (7,000 gallons per month for a family of four) and you pay on a step increase rate according to your remaining outdoor use and by lot size. The extra charge for wasting water is much steeper than Fort Collins's highest rate. Boulder's rates per thousand gallons are shown in Table 17-2. An older but indirect way to price solid waste services is the concept of recycling. If you pay by the container to dispose of solid wastes but you are allowed to send as much recycling as you like free, then the incentive to recycle is in operation.

Reducing the Cost of Government and Improving Public Services

It would be nice if all public services and goods could be privatized and the need for government would disappear, but it won't happen. In the end, some services are too public in nature, involve security, or have other attributes

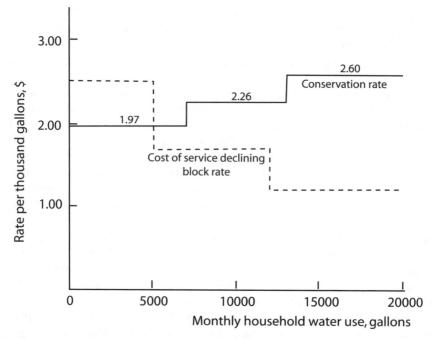

FIGURE 17-2. Comparison of conservation and cost-of-service water rates

Note: The conservation rate shown is for the City of Fort Collins, as of 2007. Fort Collins then had a base rate of $12.72, in addition to the use charges.

that require government involvement. Therefore, making government more responsive and efficient is a universally shared value.

The need for greater efficiency and effectiveness in government arises because government tends to become more inefficient as time goes along. This is simply due to human behavior, a lack of competition, and other

TABLE 17-2. Example of "water budget" rate structure for Boulder

Block	Use (% of water budget)	Rate ($)	Ratio to base rate
1	0–60	1.88	3/4
2	61–100	2.50	1
3	101–50	5.00	2
4	151–200	7.50	3
5	More than 200	12.50	5

institutional factors. When you combine this trend with the revenue crunch, government at all three levels can become unresponsive to the people it should serve. These problems cause mistrust in government, revenue problems, fiscal crisis and budget cutting, and a desire to cut taxes. They occur at a time when government action is needed for tax code reform, health care policies, and other needs for a changing population. Symptoms include people in both parties removing themselves from the political process through low voter turnouts.

These problems cascade downward in the government hierarchy. The federal government cuts programs, and then the states and local governments have problems. States have cut budgets for many programs, especially discretionary programs such as higher education and transportation. Now, new funding formulas for transportation are going to change the intergovernmental flow of revenue. Local government financial problems may seem smaller than those of the federal and state governments, but they can be significant on a case-by-case basis.

Two books offer prescriptions to improve this situation. In the first, *Reinventing Government* (Osborne and Gaebler 1992), the focus is on improving government through a mixture of measures, including the unbundling or disaggregation of services and programs to find those that can be opened to competition. In the other, *The Price of Government* (Osborne and Hutchinson 2004), the focus is on making budgeting more effective through a process called "budgeting for outcomes."

These two books explain many of the tactics that have been tried—including new political leadership, reorganizations, consolidations, and other interventions. Some help, and others are part of a list of old approaches. A joke explains some of these the issues. A new government official is at his desk pondering his future moves. He finds a note from his predecessor. She wrote: "Congratulations on your appointment. I have suggestions contained in these three envelopes, to be opened at six-week intervals."

He opens the first envelope. It says: "You need to gain control. Reorganize the agency." He does this, and it creates energetic activity, which he enjoys. After six weeks, he reads the next envelope: "Request a budget increase." This keeps everyone busy, and he cannot wait to open the third envelope. He opens it and is amazed to see that it contains three more envelopes!

The lesson is that a lot of activity occurs in government agencies, but it might be the same thing over and over again and not benefit customers, citizens, and taxpayers. This joke is not meant, of course, to denigrate the hard work of many dedicated government servants who serve the public interest. It is meant to spotlight the unproductive side of government.

In general, government has responded to these problems, and it recognizes that it does not have to be a rigid monolith that is unresponsive to public needs. There are many opportunities to make it more efficient and effective, which is the goal of the field of public administration. Programs and teaching tools abound for this purpose, such as "government executive institutes" and many training programs.

Another tool to make government more effective is to improve accountability. The budget process is supposed to create transparency and program evaluation, and the concept of the planning-programming-budgeting system is based on this assumption. However, the incentives and cultures built into the processes do not always work for public benefit.

The need for greater accountability occurs at all three levels of government. At the federal level, the Government Performance and Results Act was a 1990s statutory initiative for federal government goal setting and performance measurement (GAO 1995). After reviewing the results of this act, the Government Accountability Office found that one of the biggest problems has been achieving collaboration and that federal agencies can improve in many ways—including defining common outcomes and joint strategies, agreeing on roles and responsibilities, and improving accountability for collaborative efforts and individual performance management (GAO 2005).

The budget process at all levels of government is not always effective in improving accountability because agencies volunteer as few cuts as they think they can get away with, and they pad costs to protect against inevitable cuts. These actions are well known within the "politics of budgeting" (see Chapter 11).

Responses to these problems have led to initiatives ranging from across-the-board cuts to constitutional tax-limit proposals. These cause structural problems in the financial support web, and they may simply weaken every program equally, regardless of impact on citizens. There seems to be a need for the courage to cut antiquated and unneeded programs completely and to concentrate revenues on areas of greatest citizen need.

The incentives for government to reform itself are perverse—that is, they work at all three government levels against the changes that are needed. Newly elected political leaders would like to make marks for themselves with improvements, but they will focus on visible programs that gain votes. Government's unwieldy structure offers many opportunities for the bureaucratic hiding of inefficient programs and padding.

Citizens want value for their money, but they do not want to cut spending too much in vital public institutions. They want critical programs such as security, national defense, infrastructure, education, marketplace regulation, and many social programs. When they perceive waste, they call for tax cuts and private services, but they realize that only the public sector can provide many of the vital services.

The solution should be for public institutions to work better and smarter and to find ways around the perverse incentives of politics and the budget process to deliver the most value for taxes and fees paid. This means a combination of setting priorities and increasing efficiency through competition and market discipline. Priority setting is through the budget process, and efficiency can be improved whether services are offered by the public or private sector.

In *The Price of Government*, Osborne and Hutchinson (2004) pose five questions and challenges to shape the budget process: Is the real problem short or long term? How much are citizens willing to spend? What results do citizens want for their money? How much will the state spend to produce each of these results? And how best can that money be spent to achieve each of the core results? Consideration of these questions lead to a method for budgeting that includes getting a grip on the problem, setting the price of government that citizens are willing to pay, setting the priorities of government that citizens value most, allocating available resources across the priorities, and developing a purchasing plan for each result.

Strategies recommended include a number of the elements of the City of Charlotte's managed competition program: strategic reviews, buying services competitively, eliminating mistrust, and making administrative systems allies, not enemies. The program would include smarter work processes that include programs like total quality management and business process reengineering.

Managed Competition

"Managed competition" is a term for competition between the public and private sectors. It is a delivery system that allows either government services or privatization or a mixture of the two. For example, in the early 1990s, it changed the way that Charlotte did business and resulted in significant cost savings (Greenough et al. 1999).

The City of Charlotte (2007) has sustained its managed competition program and explains that it reviews city services for possible competition, optimization, or benchmarking with the private sector. Charlotte's approach is through a competitive bid program, which is "a planned approach for service delivery, whether the service is outsourced (no public sector competition) or private sector firms are invited to compete against the public sector for the right to provide a particular service." Charlotte has a "Privatization and Competition Advisory Committee" to help it manage its program.

Managed competition was an umbrella term for Charlotte's competitive bid program. It was to be a planned approach for service delivery, whether by outsourcing without public sector competition or competition among the private and public sectors.

Before managed competition, the Charlotte city government had a monopoly on public services, including garbage collection, landscaping, operating treatment facilities, construction, and building maintenance.

Managed competition in Charlotte has created a culture whereby government is run like a business. In 1993, the city's 26 departments were reorganized into 9 "key businesses" and 4 "support businesses." This reorganization focused on accountability for service quality and cost-efficiencies. Key businesses were required to develop business plans, and decisions about human resources, budget, finance, and purchasing, which formerly had been made by the central administrative staff, were delegated to the key business executives (who were the former department heads).

Traditional line-item budgeting was insufficient for identifying costs, making cost reductions and reengineering service delivery, so the city implemented activity-based costing, which provides the framework in which key businesses could establish full costs for activity and service levels.

Although success in Charlotte was noteworthy, acceptance of managed competition in political, community, and employee circles still has to develop. Some continue to believe that private companies have profit at heart and would gouge taxpayers, leading to inflated costs for services and lower quality. Others maintain that limits on government's ability to operate like a business lead to practices that look good on paper but do not bear up under public sector scrutiny and service delivery realities. They believe that political decisions will supersede business decisions. Some employees maintain that managed competition is a political tool to turn their jobs over to the private sector. Some are critical of the program because it disproportionately affects minorities. Unsuccessful private sector bidders argue that the city had not accurately captured all the costs of a service. They also argue that the competitive playing field favors the city government because it pays no taxes and does not have to make a profit. Other challenges include acquiring expertise to establish full costs for services, establishing credible evaluation, auditing and monitoring processes, and dedicating time to compete successfully.

Meeting Social Needs While Pricing Public Goods

The programs to make government more efficient, use pricing for public services, privatize, and deregulate all have implications for the social side of government. Though the tools of economics and finance show us how to manage infrastructure and the environment more efficiently, government also has the responsibility to consider broad social welfare. A reviewer of *The Price of Government* explained the need for this balance. He wrote that the authors "point out that while much of what is discussed in the book could be summed up under the category of market-oriented government, markets are only half the answer." They recognize that "markets are impersonal, unforgiving, and, even under the most structured circumstances, inequitable. . . . They conclude that entrepreneurial governments must embrace both markets and community as they begin to shift away from administrative bureaucracies" (London 2007).

Applications of New Tools to Infrastructure and Environment

This final chapter gives us a focus on how to balance government and private sector activities in managing infrastructure. It explains complex and evolving phenomena that include deregulation and privatization, along with the necessary tools such as the unbundling of services, managed competition, and pricing public goods to manage demand. The outcomes of these management innovations include reducing the cost of government, improving public services, and meeting social needs while allocating public goods more responsibly.

References

Brown, A. C. (1983). "For sale: Pieces of the public sector." *Fortune*, October, 31.

Butler, Stephan. (2007). "Turning public works into private ventures will revitalize the U.S." *Engineering News-Record*, July 4.

City of Boulder. (2007). The basics of your water budget. http://www.boulder colorado.gov. Accessed September 4, 2007.

City of Charlotte. (2007). http://www.charmeck.org/. Accessed September 4, 2007.

Cornell University. (2007). Governmental restructuring. http://government. cce.cornell.edu/government_restructuring.asp. Accessed March 8, 2007.

GAO (U.S. Government Accountability Office). (1995). *Managing for results: Status of the Government Performance and Results Act*. Report T-GGD-95-193. U.S. Government Printing Office, Washington, DC.

———. (2005). *Results-oriented government: Practices that can help enhance and sustain collaboration among federal agencies*. Report GAO-06-15. U.S. Government Printing Office, Washington, DC.

Greenough, G., Eggum, T., Ford, U., Grigg, N., and Sizer, E. (1999). "Public works delivery systems in North America: Private and public approaches, including managed competition." *Journal of Public Works Planning and Management*, 4(1), 41–49.

London, S. (2007). Reinventing government: Book review. http://www. scottlondon.com. Accessed September 4, 2007.

Osborne, D., and Hutchinson, P. (2004). *The price of government: Getting the results we need in an age of permanent fiscal crisis*. Basic Books, New York.

Osborne, D., and Gaebler, T. (1992). *Reinventing government: How the entrepreneurial spirit is transforming the public sector.* Addison-Wesley, Reading, MA.

Walker, D. L. (1985). "The economics of privatisation." Paper presented at International Water Pollution Centre symposium on privatization and the water industry, London, March.

Index

About the Author

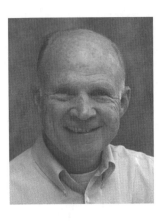

Neil S. Grigg, Ph.D., P.E., D.WRE, is a professor and former head of the Department of Civil and Environmental Engineering at Colorado State University. His professional work is focused on infrastructure and environmental management, with an emphasis on water, transportation, and utilities. He graduated from the U.S. Military Academy at West Point and has graduate degrees from Auburn University and Colorado State University. He participates actively in research and policy studies about infrastructure and public works issues, and he has been on the Board of Directors of the American Public Works Association and the National Research Council's Board on Infrastructure and the Constructed Environment. His teaching focuses on courses on infrastructure management and security and on water resources planning and management, including economics and finance content. He is the author of several textbooks and a large number of articles about infrastructure and environmental systems, with a focus on water resources. He teaches a P.E. refresher course on engineering economics. His experience includes a number of consulting assignments, and he is cofounder of a Denver-area consulting firm. As assistant secretary for natural resources for North Carolina, he had responsibility for several state natural resources agencies, and he was an environmental regulator for the state. He has also been director of two state water resources institutes and a consultant to other state governments. He has been a member of the water and transportation boards of the City of Fort Collins. His international experiences include service as a policy consultant in several countries. He has served in a U.S. Supreme Court appointment as the river master of the Pecos River since 1988.